Felix Müller

Formation of Spatio-temporal Patterns in Stochastic Nonlinear Systems

Felix Müller

Formation of Spatio-temporal Patterns in Stochastic Nonlinear Systems

A study of extended systems with reference to abstract biological mechanisms

Südwestdeutscher Verlag für Hochschulschriften

Impressum / Imprint

Bibliografische Information der Deutschen Nationalbibliothek: Die Deutsche Nationalbibliothek verzeichnet diese Publikation in der Deutschen Nationalbibliografie; detaillierte bibliografische Daten sind im Internet über http://dnb.d-nb.de abrufbar.

Alle in diesem Buch genannten Marken und Produktnamen unterliegen warenzeichen-, marken- oder patentrechtlichem Schutz bzw. sind Warenzeichen oder eingetragene Warenzeichen der jeweiligen Inhaber. Die Wiedergabe von Marken, Produktnamen, Gebrauchsnamen, Handelsnamen, Warenbezeichnungen u.s.w. in diesem Werk berechtigt auch ohne besondere Kennzeichnung nicht zu der Annahme, dass solche Namen im Sinne der Warenzeichen- und Markenschutzgesetzgebung als frei zu betrachten wären und daher von jedermann benutzt werden dürften.

Bibliographic information published by the Deutsche Nationalbibliothek: The Deutsche Nationalbibliothek lists this publication in the Deutsche Nationalbibliografie; detailed bibliographic data are available in the Internet at http://dnb.d-nb.de.

Any brand names and product names mentioned in this book are subject to trademark, brand or patent protection and are trademarks or registered trademarks of their respective holders. The use of brand names, product names, common names, trade names, product descriptions etc. even without a particular marking in this works is in no way to be construed to mean that such names may be regarded as unrestricted in respect of trademark and brand protection legislation and could thus be used by anyone.

Coverbild / Cover image: www.ingimage.com

Verlag / Publisher:
Südwestdeutscher Verlag für Hochschulschriften
ist ein Imprint der / is a trademark of
AV Akademikerverlag GmbH & Co. KG
Heinrich-Böcking-Str. 6-8, 66121 Saarbrücken, Deutschland / Germany
Email: info@svh-verlag.de

Herstellung: siehe letzte Seite /
Printed at: see last page
ISBN: 978-3-8381-3420-8

Zugl. / Approved by: Berlin, HU, Diss., 2011

Copyright © 2012 AV Akademikerverlag GmbH & Co. KG
Alle Rechte vorbehalten. / All rights reserved. Saarbrücken 2012

Zusammenfassung

Die vorliegende Arbeit befasst sich mit einer Reihe von Fragestellungen, die Forschungsfeldern wie rauschinduziertem Verhalten, Strukturbildung in aktiven Medien und Synchronisation nichlinearer Oszillatoren erwachsen. Der Fokus liegt auf der mathematischen Modellierung komplexer Mechanismen. Dabei verfügen die verwendeten nichtlinearen Modelle über erregbare, oszillatorische und bistabile Eigenschaften als elementare Moden dynamischer Systeme. Das Zusammenwirken mit stochastischen Fluktuationen trägt wesentlich zur Entstehung komplexer Dynamik bei.

Zu Beginn werden mittlere Verweilzeiten im stationären Zustand des FitzHugh-Nagumo (FHN) Modells unter dem Einfluß stochastischer Fluktuationen untersucht. Das Koppeln mehrerer solcher erregbaren Einheiten führt zur Differenzierung zweier Erregbarkeitstypen, von denen nur ein Typ zusammenhängende Keimbildung zulässt.

Im folgenden Kapitel wird qualitativ untersucht, auf welche Weise sich die extrazelluläre Kaliumkonzentration, die von umliegenden Neuronen gespeist wird, auf die Aktivität dieser Neuronen auswirkt. Als Basismodell wird das FHN Modell bemüht, das um eine Gleichung für die extrazelluläre Konzentration erweitert wird. Neben der Untersuchung lokaler Aktivität wird vor allem die Ausbildung ausgedehnter Strukturen in einem heterogenem Medium analysiert. Die raum–zeitlichen Muster umfassen Wellenfronten und Spiralen aber auch ungewöhnliche Strukturen, wie erratisch wandernde Cluster oder invertierte Wellen.

Eine wesentliche Rolle bei der Ausprägung von solchen raum–zeitlichen Strukturen spielen die Randbedingungen des betrachteten Systems. Die Beschreibung von bistabilen Fronten und deren Wechselwirkungen mit Dirichlet Randbedingungen ist der Hauptaspekt im 4. Kapitel. Sowohl für diskret gekoppelte bistabile Elemente als auch für kontinuierliche Fronten werden Methoden zur approximativen Berechnung von Frontgeschwindigkeiten vorgestellt. Typische Bifurkationen werden quantifiziert und diskutiert.

Der Rückkopplungsmechanismus aus dem Modell zur Beschreibung von neuronalen Einheiten und deren passiver Umgebung kann weiter abstrahiert werden. Hauptaugenmerk des 5. Kapitels ist die Behandlung eines Zweizustandsmodells, das über zwei Wartezeitverteilungen definiert wird, welche erregbares Verhalten widerspiegeln. Die Rückkopplung besteht aus der mittlere Gesamtaktivität eines Ensembles, das die individuellen Erregungszeiten beeinflußt. Untersucht wird die instantane und die zeitverzögerte Reaktion des Ensembles auf diese Rückkopplung. Im Fall von zeitverzögerter Rückkopplung tritt eine Hopf–Bifurkation auf, die zu stabilen Oszillationen der mittleren Gesamtaktivität führt.

Das letzte Kapitel befasst sich mit Diffusion und Transport von Brownschen Teilchen in einem raum–zeitlich periodischen Potential. Auch hier sind es Synchronisationsmechanismen, die im Falle eines konstant gekippten Potentials nahezu streuungsfreien Transport ermöglichen. Für die symmetrische Situation, beleuchten wir die Ursache einer erhöhten effektiven Diffusion und gelangen zu einer quantitativen Abschätzung der maximierenden Parameter.

Resümierend wird die Wirkung von nichtlinearer Dynamik und stochastischen Fluktuationen auf räumlich ausgedehnte und gekoppelte Systeme untersucht. Die Entstehung von kohärenten Strukturen, sei es in Raum oder Zeit, führen wir auf Übergänge zwischen dynamischen Zuständen und auf Synchronisationseffekte zurück.

Contents

1	**Preface**	**1**
1.1	Excitable systems	6
1.2	Neuron models	6
1.3	Reaction–diffusion systems	11
2	**Stochastic escape from a fixed point**	**15**
2.1	Origin of stochastic fluctuations	15
2.2	Escape from a parabolic potential	17
2.3	Mean escape time in the FHN dynamics	20
	2.3.1 Probability density distribution in phase space	20
	2.3.2 Complete time scale separation	22
	2.3.3 Spatially extended excitable and sub–excitable medium	23
2.4	Conclusions	29
3	**Dynamical structures in a heterogeneous active medium**	**31**
3.1	Introduction	31
3.2	Abstract model for a potassium–driven neuron	33
	3.2.1 Biological background	33
	3.2.2 Local model	34
	3.2.3 Fixed points and bifurcations	36
3.3	Local dynamics under the influence of noise	40
3.4	Two excitable units interacting with a common exterior	44
3.5	Patterns in a spatially extended medium	46
	3.5.1 Waves, spots and spirals (Fig. 3.10)	48
	3.5.2 Noise supported traveling clusters (Fig. 3.11)	49
	3.5.3 Desynchronized oscillators embedded in a z-sea (Fig. 3.12)	50
	3.5.4 Oscillations form a propagating ring-like pattern (Fig. 3.13)	51
	3.5.5 Bistability and inverted waves. (Fig. 3.15)	53
3.6	Noise induced regularity at absorbing boundaries	54
3.7	Conclusions	56
4	**Bistable wave fronts interacting with boundaries**	**59**
4.1	Introduction	59
4.2	Local kinetics and linearization	63

Contents

    4.3 Array of coupled units 64
            4.3.1 Thin front 64
            4.3.2 Propagation failure 67
            4.3.3 Interaction with the boundaries 68
            4.3.4 Self–sustained oscillation near the boundary 70
    4.4 Continuous bistable front 72
            4.4.1 Diffusing activator – immobile inhibitor 73
            4.4.2 Activator and inhibitor diffusion 82
    4.5 Conclusions 88

5 Excitable two–state units coupled with delayed feedback   91
    5.1 Introduction 91
    5.2 Individual unit 92
            5.2.1 Model definition 92
            5.2.2 Interspike interval distribution and power spectral density . . 93
            5.2.3 Generalized master equations 95
    5.3 Ensembles of globally coupled units 96
            5.3.1 Instantaneous coupling 96
            5.3.2 Delayed coupling 100
    5.4 Numerical simulations 105
    5.5 Conclusions 109

6 Synchronization and transport in an oscillating periodic potential   111
    6.1 Introduction 111
    6.2 Theta neuron 112
    6.3 Dynamical equation 114
    6.4 Effective diffusion in case of $F = 0$ 116
    6.5 Synchronization in a biased potential 121
    6.6 Stationary probability density in the $y-$ plane 126
    6.7 Conclusions 130

7 Summary   133

Appendix A   137
    1   Mean first passage time 137

Appendix B   139
    2   Stationary solutions in the globally coupled two–state system 139

1 Preface

Let us imagine a purely linear world where the laws of nature would be limited to two types of elementary mechanics. First exponential growth or decay mathematically captured by a first order differential equation $\dot{x} \propto \pm x$. The second type are harmonic oscillations described by a second order differential equation $\ddot{x} \propto -x$. If matter would follow these laws exclusively the universe would be a bleak and bland spot. Fortunately for us living creatures there are ruling principles of nature beyond such linear laws that enable complex interaction between physical objects which finally leads to genesis of life.

Basically one could state that every kind of natural science has its own specific approach to understand the very same secret; the becoming of life and humankind out of the elementary components of matter. Yet sixty years ago the hierarchy of the natural science was clear. The principal constituents of non–living being such as energy, matter and the laws of their interaction were the topic of physics. The reaction of inorganic atoms and molecules up to organic structures was studied by chemists. The classification of living things, microbes, plants, animals and the analysis of their composition was reserved for biology and, maybe last, psychologists concentrated on the complex occurrence the human (sub–)consciousness produces. Nowadays the borderlines between these sciences are blurred. Physicists are working together with psychologists in order to bring to light how the human brain makes us to thinking creatures. Researchers in theoretical biology use methods from fluid mechanics to find out how microorganisms move in liquids, and so on.

The research field of stochastic processes and nonlinear dynamics in which the present work is embedded, can be considered to be close to theoretical concepts of biology. Nevertheless, it is also a field of pure physics with the ambition to understand the fundamental principles of complexity. Hence, we will work with models that indeed have a biological background. On the other hand, these models are often too simplified to be able to give quantitative predictions for real biological situations. However, they are provided to establish the dynamical essence behind complex processes. There are three basic dynamical regimes that are produced by such simplified and nonlinear models, which are excitability, self–sustained oscillations and bistability. Their dynamical characteristics and transitions between them through bifurcations are a main aspect in this work.

Nonlinear systems that exhibit such dynamical regimes are abstract model systems to describe phenomena which we run across in every day life. In fact, every event triggered by a strong enough external impact, which evokes a process of subsequent

1 Preface

events creating a self–energizing course of action, is a nonlinear response known as positive feedback.

(i) In case of excitable systems intrinsic slowdown mechanisms inhibit further activation and drive the system back into the original state. A typical representative showing excitable behavior are neuron cells [21]. However, this concept can be successfully applied to completely different fields such as economics [135]. Especially when the system is spatially extended and excitable waves are formed it is a powerful method even to describe the development of urban crime [131].

(ii) For self–sustained oscillations there is no stable rest state and the slowdown (or inhibiting) process crosses over into the self–feeding activation again. This procedure is repeated recurrently and it is stable against external perturbations. It is the basic framework to explain synchronization effects which are based on stable and self–sustained periodic sequences. Famous and popular examples of such self–sustained oscillating behavior are the rhythmic blinking of fireflies from south–east Asia [15] or the oscillating number of preys and predators described by A. Lotka and V. Volterra [92, 155].

(iii) Bistability as the third kind of elementary nonlinear dynamics is interesting under the aspect of decision processes. In linear systems transitions are always smooth and do not lead to a qualitative difference. In bistable systems the two stable steady states separated by a barrier or threshold can be associated with two distinct scenarios between which the system can transit. Coexistence of two stable phases is a generic property for a widely spread class of systems. The classical example from physics is the mechanism of ferromagnetism and the model for the alignment of the electron spins by Ising [57]. If the transition from one state to another propagates in space forcing the system into the new regime we refer to this as a bistable propagating front, sometimes also called Bloch–fronts in literature [45]. For chemical reactions F. Schlögl obtained a dynamical equation with a cubic nonlinearity that possesses two stable phases crossing over via a second order phase transition [127]. Surprisingly, the spread of the Black Death over Europe in the 14th century can be explained by assuming a simple nonlinear model that produces a bistable front [99]. It even yields a good approximation of its propagation velocity. Modern research about the spread of epidemics, however, requires extended models and concepts beyond diffusion [49].

In addition to nonlinearity and spatial diffusion we study a third significant factor of complexity that influences the physical world and especially any biological process. We will discuss the effects of stochasticity or noise, present in any system at finite temperature as explained in the next chapter. In the models we investigate in this work, noise often plays the role of the driving mechanism that forces a system out of stable state. Concerning the previously mentioned bistable systems the impact of noise can lead to the famous effect of stochastic resonance [31], an evidence that the presence of noise can have a constructive impact.

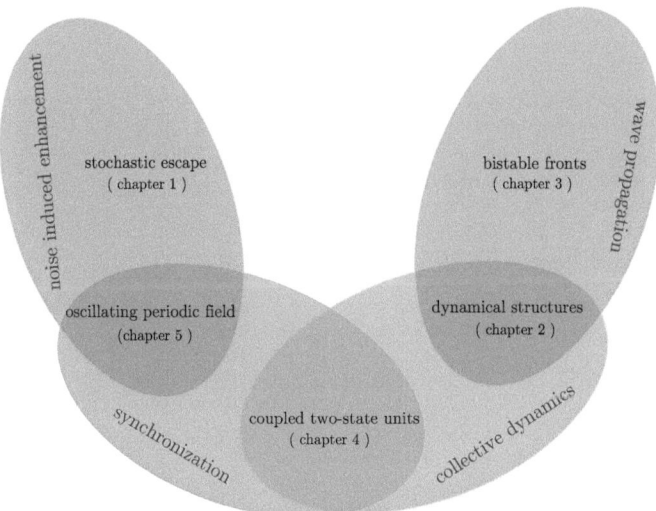

Figure 1.1: Schematic overview of chapters of the thesis and their thematic interrelation beyond the numeric order they appear.

In the following we give a brief outlook over the central themes of the thesis. After the introduction (chapter 1) we begin with a consideration of the influence of noise on excitable systems in chapter 2 – **Stochastic escape from a fixed point**. We estimate the mean escape time out of a parabolic potential arriving at Kramers' famous formula and extend this examination on single and coupled excitable units. The basic model we use is the FitzHugh–Nagumo (FHN) model, also in the following chapter 3 – **Dynamical structures in a heterogeneous active medium**. Here we study excitable units embedded in an inactive environment modeled on the basis of ion exchange mechanisms from cells into the extracellular space by introducing an extra dynamical equation. The resulting local dynamics possesses each of the three mentioned regimes; excitability, oscillatory behavior and bistability and even their coexistence. Different patterns partly well known but also extraordinary structures appear for the spatially extended and heterogeneous system under the variation of

1 Preface

external parameters. Which type of patterns is created and its lifetime depends significantly on boundary conditions. This is the topic of chapter 4 – **Bistable wave fronts interacting with boundaries**, where we study bistable fronts and their interaction with Dirichlet boundaries by analyzing the front velocity. Far away from the boundaries we obtain approximated expressions for the bulk velocity that turn out to be mono– or bistable. Moreover we find a relation between the distance of the front to the boundary and the velocity that gives us the possibility to predict roughly which interaction with the boundary takes place. In chapter 5 – **Excitable two–state units coupled with delayed feedback** we take up the idea from chapter 3 where the additional dynamics induce a positive feedback mechanism. On the basis of waiting time distributions we define coupled excitable units in which the mean activity of the whole ensembles is fed back instantaneously and later also delayed. We find monostable and bistable regimes for the common output. For the case of delay even oscillatory behavior can be obtained where the units are synchronized. The phenomenon of synchronization is again an important aspect in chapter 6 – **Synchronization and transport in an oscillating periodic potential**. A spatially periodic potential with oscillating amplitudes is considered which implies properties of ratchets, phase oscillators and excitable dynamics. We investigate directed transport and diffusion for a finite tilt and correlated synchronization effects. For an unbiased potential we describe the interaction of noise and the oscillating amplitudes which can lead to a fast diffusive spread within the system. The probability density distribution in the corresponding phase space is determined to shed light on underlying dynamical mechanisms.

Parallel to the numeric order the chapters appear in the thesis they are related crossways on the level of common subject areas. This is schematically depicted in Fig. 1.1. Spatially extended propagating waves are the main subject of the chapters 2 & 3, the interplay of interacting active units and their collective behavior is an important aspect in the chapters 2 & 4, individual oscillations leading to synchronization effects are treated in the chapters 4 & 5 and finally, effects of noise induced phenomena are discussed in the 1st & 5th chapter.

As a general aim of the work we want to demonstrate that diverse fields connected to complex behavior ranging from pattern formation to synchronization can be analyzed by models which have a relatively simple structure and no more than the three mentioned dynamical regimes. With such abstract models we can learn more about the general properties of the dynamical regimes beyond linear response. That comes along with the discovery of fundamental bifurcation scenarios that we find for different quantities. We can generate complex patterns which correspond qualitatively to biological situations and find synchronization in two distinct systems, both by applying relatively simple model components. For each model system we can identify the dominant control parameters and show their impact on the dynamics.

They constrain the system into the dynamical regimes and determine whether the output is synchronized, a front rebounds at a boundary or the effective diffusion is maximized. Knowledge about the influence of these governing parameters gives us finally control over the dynamical behavior of such simplified systems. However, it can also enable us to make predictions for more complex systems.

1 Preface

1.1 Excitable systems

Excitable systems are a subclass of dynamical systems, i.e. they are described by a set of variables and rules or functions that define their evolution in time. The space spanned by these variables is called phase space. In phase space the dynamics of such systems is portrayed as trajectories, which are the parametric curves displaying the variation of the dynamical variables and their evolution in time.

Characteristic features of the structure of phase space identify the class of excitable systems. They have one linear stable fixed point as a steady state with its belonging basin of attraction. This basin defines the zone of influence of the stable fixed point. Starting from initial conditions within this basin of attraction, trajectories decay to the steady state, if the corresponding eigenvalues are real. If they have an imaginary part, trajectories spiral to the steady state. Outside the basin of attraction trajectories first run away from the equilibrium point and perform a characteristic large excursion in phase space before they eventually come back to end in the steady state. So, the region beyond the mentioned basin of attraction finally brings trajectories back to the steady state and should be also considered as a kind of a larger basin of attraction. In order to distinguish the inner and outer region the demarcation between them is called threshold. This threshold can be either smooth or hard steplike. For the first case trajectories starting within the threshold zone exhibit an amplitude of their excursion which increases continuously with the distance of their initial values to the fixed point. This is known as canard–explosion. The hard threshold is a defined subspace and separates initial conditions whose trajectories either directly lead back to the steady state or let them run away.

Besides the existence of a steady state and a threshold the response on perturbations is the third important feature. Perturbing an excitable system in the steady state either results in sub– or superthreshold behavior. That means elongating the system with a perturbation below a critical value it immediately relaxes back into the steady state, however, when the critical perturbation is exceeded the system is forced beyond the threshold followed by the long journey through the phase space. During the latter further perturbations do not change the dynamics drastically.

1.2 Neuron models

Typical representatives of excitable systems in mathematical biology or biophysics are neuron models. There is an immense variety of such models and a wide research field around them describing the biological background more or less precisely [60, 40].

Neurons are cells that transport information coded as electro–chemical signals. Their membranes separate intra– and extracellular ion concentrations. For example the potassium ion concentration within the cells is much higher than in the extracellular space. This is treated in more detail in chapter 3. To prevent these charged

1.2 Neuron models

ions from balancing through the membrane by diffusion a membrane potential is needed that creates an equilibrated situation. This potential can be derived with the Nernst equation [60]

$$V_{eq} = \frac{RT}{ZF} \ln \frac{[I]_{\text{out}}}{[I]_{\text{in}}}. \tag{1.1}$$

The argument in the logarithm is the fraction of outer and inner ion concentration; R is the universal gas constant; T is the temperature (typically $T = 310°K$ for living mammals); F is Faraday's constant and Z is the valence of the ion. Note that this equation assumes only one type of ions producing the concentrations $[I]$. In a real biological scenario a number of different ions are involved in neuronal processes. However, in most cases it is sufficient to consider the governing ions sodium, potassium, calcium and chloride (in the following denoted as Na^+, K^+, Ca^{2+} and Cl^-), each of them has its own membrane potential. In some models the transport of remaining ions is subsumed as a leak current. The typical common membrane potential in the equilibrium is about -70 mV [40] and is also called resting potential. It corresponds to the steady state and with that we have the first ingredient for an excitable system.

In 1952 in their experiment with squid axons Hodgkin and Huxley discovered the basic mechanisms of voltage and gating dynamics [54]. The pioneering Hodgkin-Huxley (HH) model consists of one equation for the evolution of the membrane potential including K^+-, Na^+- and a leak current and three equations for the gating variables, which control the opening and closing of ion channels through the membrane. The latter are in their common effect inhibiting processes whereas the voltage dynamics is the activator process. We will not write the four equations here, since they will not be further treated in this work.

Instead we are going to have a closer look at simplified neuron models, that have a reduced number of equations and manageable set of parameters. It is clear that the reduction to less number of variables leads to a minimization of available dynamical regimes. For instance less equations come along with fewer participating time scales and thus the most two component models cannot provide the transition from single spiking to tonic bursting behavior.

The working range of parameters controlling the dynamical regime of the models is mostly chosen to be close to bifurcations, so that the system response is sensitive to perturbations. Considering the form of the bifurcation and the properties after the transition two excitability types of neuron models can be distinguished:

Excitability type I:

The excitability of type I characterizes neuron models that undergo a global bifurcation such as 'saddle–node–infinite–period' (SNIPER) bifurcation at a critical value of the control parameter. Beyond this critical value the steady state is unstable and is surrounded by a stable limit cycle. The region of excitability is left

1 Preface

by a discontinuous increase of amplitude and a soft square-root-like growth of the oscillation frequency starting from zero level, corresponding to an infinite period time. Examples for neuron models showing type I excitability are the Ermentrout– or the Izhikevich model [122, 60].

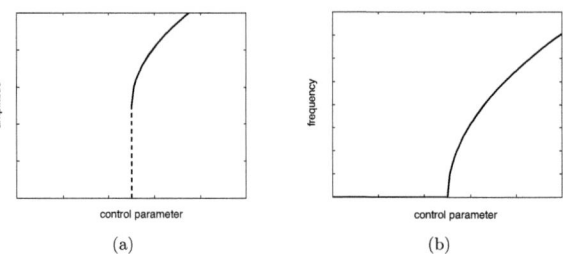

Figure 1.2: Illustration of amplitude (a) and frequency (b) behavior close to bifurcation for type I excitability

Excitability type II:
In the case of excitability II, the dynamical system exhibits a Hopf–bifurcation due to the tuning of a control parameter. Then the amplitude and frequency behave inversely to type I. Oscillations after the bifurcation start with a soft square-root like amplitude and with an immediate finite frequency. Examples for neuron models showing type II excitability are the Hodgkin–Huxley– or the Morris–Lecar model [54, 96]. Both types of behavior were observed in measurements already in 1948 by

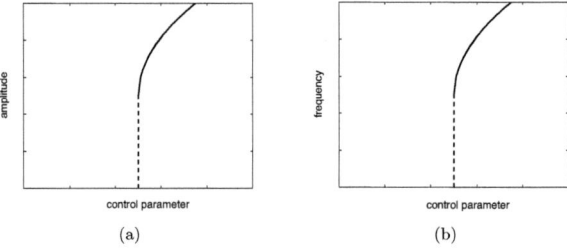

Figure 1.3: Illustration of amplitude (a) and frequency (b) behavior close to bifurcation for type II excitability

Hodgkin, however the description with bifurcations followed decades later in 1989

by Rinzel and Ermentrout [122].

In the following sections a two–component neuron model, the FitzHugh–Nagumo model (FHN) [34] is often used as a canonical model that exhibit excitable or bistable behavior. Originally it was introduced as a simplification of the HH model to design the essential properties of spike generation of sodium and potassium ion flow. One possible representation reads

$$\dot{u} = u - \frac{1}{3}u^3 - v + I, \quad \dot{v} = (au - v + b), \qquad (1.2)$$

where u denotes the dimensionless voltage variable, describing the activator with its significant cubic nonlinearity and v stands for the recovery variable, that inhibits the dynamics. The time scales of both processes are separated by which is chosen to be small. This ensures a fast activator and slow inhibitor dynamics. The system can be stimulated by an external signal I perturbating the system sub–or superthreshold, when the system is in the excitable regime. Regarding the nullclines of the dynamics, where $\dot{u} = 0$ for the activator and $\dot{v} = 0$ for the inhibitor, the different dynamical regimes are determined by their relative position. The FHN model approximates dynamics from models closer to specific biological situations with the cubic shaped nullcline and a linear inhibitor nullcline. However, there is also an theoretical correspondence to a chemical reaction scheme, given in [93].

In Fig. 1.4 (a,c,d) the three existent dynamical regimes are illustrated in phase space where the thin black lines show the nullclines. The arrows stand for the vectorfield of Eqs. 1.2 and a sample trajectory for each case is drawn with thick solid lines. The regimes are distinguished by the choice of inhibitor parameters a and b, that moves the nullcline position while the cubic nullcline remain fixed. Thus the nullclines have either one or three intersections according to fixed points of the dynamics ($\dot{u} = \dot{v} = 0$) which can be stable or unstable.

Fig. 1.4 (a) represents the excitable regime including one stable fixed point with the corresponding pulse shape in time, shown in Fig. 1.4 (b) for the activator (solid) and the inhibitor (dashed) versus time. In order to describe the depolarization of a neuron with a rest state at a negative potential the typical excitable FHN model has its fixed point at negative u–values. Initiated far enough from that fixed point trajectories moves to the right stable nullcline branch for which the neuron is conceived to be active. After a certain engaged time it get attracted by the left stable cubic branch and passes the refractory period before the entire excursion in phase space is completed. Each of the mentioned phases is related to a part of the depolarization spike in time. However, for a fixed point position close to the maximum of the cubic nullcline the system acts as a excitable dynamics, too. The spikes then turn into polarization spikes and the whole situation would be inverted. We will confront ourselves with such a scenario in chapter 3.

Fig. 1.4 (c) shows the bistable situation with two stable fixed points and one

1 Preface

unstable fixed point in between. Trajectories started in the associated basin of attraction end either in the left or in the right stable fixed point. This regime is considered more precisely in chapter 4. The remaining figure (d) represents the oscillatory regime with only one unstable fixed point surrounded by a stable limit cycle. Since this is the only attractor all trajectories move along a periodic orbit circulating around the fixed point performing self–sustained oscillations in time. Finally they end on that limit cycle for $t \to \infty$.

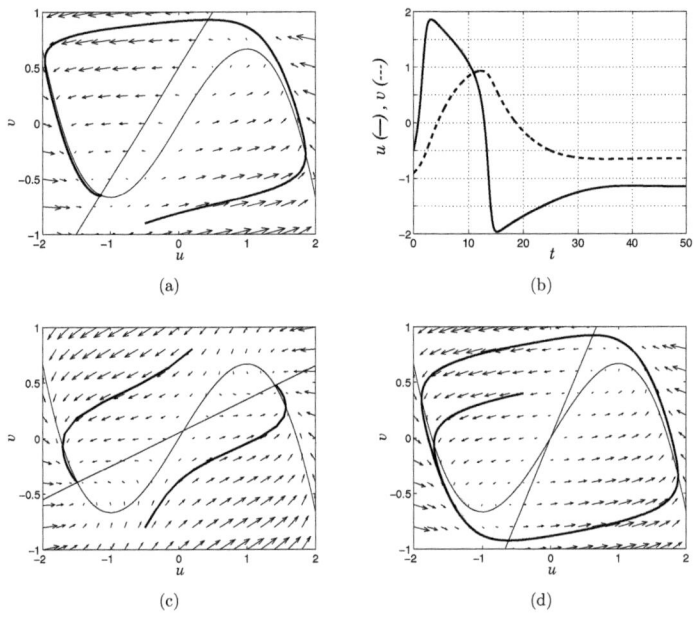

Figure 1.4: Three dynamical types of behavior provided by the FHN model. (a) and (b): excitable regime with the corresponding pulse shape in time, solid line: activator, dashed line: inhibitor, (c) bistable and (d) oscillatory regime. Arrows present the direction field of Eqs. 1.2.

1.3 Reaction–diffusion systems

Equations are called reaction–diffusion systems (RDS) if they can be divided into a locally acting reaction and a spatially distributing diffusion part, as the name might suggest. In mathematical terms reaction–diffusion systems are parabolic equations and describe the temporal and spatial evolution of a density field. They can be also used as a macroscopic representation of microscopic processes in which a large number of particles or individuals move randomly and thus spread in space. By a transition to a continuum and interpreting the ensemble of single particles as a concentration or density field its motion can be identified as a diffusion process. Intrinsic interactions that might bias the collective dynamics find reflexion in an additional reaction term.

Reaction–diffusion systems serve as abstract models for pattern formation such as moving waves, spirals, spots, target patterns, Turing patterns and many more. Besides pattern formation in physical systems they are of great relevance in chemistry, biology and even medicine. To exemplify we want to point out three examples.

One of the origins of research on this field is the investigation of the famous Belousov–Zhabotinsky (BZ) reaction. Well stirred this is a chemical reaction whose products oscillate in time. Fortunately these products have different color and the oscillation period is in the order of seconds. Therefore the oscillations can be seen with the naked eye. If the reaction is ignited at a local spot and remains unstirred within a petri dish, recurrent ring–like waves appear which propagate away from the initial point [166]. These wave can break under certain intrinsic circumstances or caused by external stimuli which leads to free wave ends. Their planar form is unstable and they begin to curl forming the source of a spiral–wave [1]. For the BZ reaction there is a model that consists of three variables describing the most important chemical agents involved. This model that mimics quantitatively the chemical reaction is known as the Oregonator model [99]. The dynamics of spiral–waves can be controlled in the light–sensitive BZ reaction and dynamical equations for the tip motion can be derived [133, 97].

Spiral waves in biological systems have been found in experiments with xenopus laevis oocytes, which are egg cells from the African clawed frog [81]. Within such cells variations in calcium ion concentration are induced by a trigger whereupon chemical concentration waves and spirals appear, made visible by Ca^{2+} dyes. The dynamics of calcium propagation as a significant signaling mechanism has been modelled on various abstraction levels [165, 83, 29].

Observed electrophysiological patterns in heart tissue of mammals are also a biological topic with strong relation and high impact on medical science. It has been suggested, that ventricular fibrillation, which is a cardiac arrhythmia leading to sudden cardiac death in many cases, originates in a disturbance of the regular self–sustained electrical scroll waves that cause the heart beat. They have been verified in heart tissue of dogs or rabbits [158, 25] Understanding the underlying dynam-

1 Preface

ics or the reason of this disturbance could help to invent supression techniques for ventricular fibrillation.

Each of the presented examples of pattern formation and many more originate from rather different physical situations. Nevertheless, the phenomenology and mesoscopic rules of appearance and motion reflect generic features of such systems. Essential conditions have to be fullfilled such as being far from equilibrium and thus driving by energy dissipation and pumping. Therefore those pattern are often called dissipative structures.

Quantitative description serve reaction–diffusion equations generally written as

$$v_i(x,t) = f_i(v_1...v_n) + D_i \Delta v_i, \quad i = 1...n, \tag{1.3}$$

where v_i is the i-th field of a set of n contributing fields. For instance it stands for a concentration of a chemical substance when a chemical reaction shall be described. This concentration is associated with a diffusion coefficient D_i which is assumed to be constant here and for the whole presented work. The reaction kinetics is described by the function f_i that can include the dependencies on every substance that participates in the reaction. So, the field propagates in space and time, driven by local reaction and diffusive dispersion.

The presented examples demand reaction functions that describe excitable behavior as presented in the previous section. These reaction terms do not need to have their origins in neuron models. For chemical reactions they can be derived from the reaction kinetics as mentioned for the Oregonator model describing the BZ reaction. To refer such media in a more general manner it is often called active media.

In this work we will often apply the FHN dynamics from Eqs. 1.2 as reaction term. Due to the clear nullcline structure it is a transparent model and thus convenient for investigating pattern formation phenomena. In Fig. 1.5 a snapshot of a doubled spiral is shown, simulated with a spatial extended FHN model. Characteristic behavior of FHN wave fronts in one and two dimensions as well as the influence of stochastic fluctuations on them has attracted interest [45, 100]. To give the last tersely examples waves and spirals found in the famous catalytic CO oxidation on Pt(110) and in the completely different scope of the slime mould Dictyostelium discoideum has been described by the spatial extended FHN dynamics [6, 152].

1.3 Reaction–diusion systems

Figure 1.5: Spiral patterns due to a spatial extended FHN system. Grey level refer to the different activator values. (Colors online)

2 Stochastic escape from a fixed point

2.1 Origin of stochastic fluctuations

As mentioned in section 1.2 the activation of excitable systems is caused by external perturbation. Before we discuss the origin of these perturbation we give a brief overview of the physical source of stochastic fluctuations.

In 1829 Robert Brown reported about jittering pollen when he looks at water through a microscope and thought he found the microscopic origin of life [14]. In fact, he was not the first one, who observed this erratic motion of little things in fluids, however, since his publication this effect is known as Brownian motion. Although his conclusion about the living nature of his observations was a misinterpretation, the stochastic force he found is indispensable for e.g. neuronal functioning and therefore for the functioning of living being. Also macroscopic and directed motion of creatures is driven by noise. Almost 80 years later Paul Langevin presented a method how stochastic forces can be mathematically described on the basis of Newton's equations of motion [79]. The following short consideration will make clear, that Brownian motion has a purely physical background and is finally caused be the surrounding temperature.

The dynamics of a free particle with the mass m and radius R_p moving in a fluid with the viscosity   is given for simplicity in one spatial dimension by

$$m\ddot{x} = -6R_p \dot{x} + F(t). \qquad (2.1)$$

The knowledge of the discontinuous nature of fluids on molecular scale is reflected in the force $F(t)$. This said molecules, the fluid consists of, are assumed to have a much smaller mass and kick the considered particle in infinitesimal short time at distinct times t_i which causes the jitter motion observed by Brown. The responsible stochastic force may be written as $F(t) = \sum_i f_i\,(t - t_i)$ with instantaneous kick strength f_i at time t_i. In average the force acts equally in each direction yielding $\langle F(t) \rangle = 0$. Writing the Stokes' friction as $\Gamma = 6r$ and averaging the dynamical equation 2.1 leads to

$$m\frac{d}{dt}\langle x\dot{x}\rangle = -\Gamma\langle x\dot{x}\rangle + m\langle \dot{x}^2\rangle + \langle xF(t)\rangle. \qquad (2.2)$$

The last averaged term contains two independent variables and vanishes. Now, we need to put a strong condition. We assume the system to be in thermal equilibrium so that, following the equipartition theorem, every degree of freedom possesses the same amount of energy and we can identify

$$\frac{1}{2}m\langle \dot{x}^2\rangle = \frac{1}{2}k_B Q, \qquad (2.3)$$

with the temperature Q of the system and k_B as the Boltzmann constant. We can substitute the latter expression with the second term on the right hand side of Eq. 2.2 and integrate subsequently assuming that the particle started in $x = 0$ at time $t_0 = 0$. After neglecting short time behavior directly after a fluid molecule hit the particle we arrive at

$$\langle x^2 \rangle = 2\frac{k_B Q}{\Gamma}t. \qquad (2.4)$$

So, the covered mean distance of a particle that is randomly hit by fluid molecules growth proportional to $\sqrt{t}$ in contrast to a deterministic particle performing ballistic motion whose proportionality reads $x(t) \propto t$. Finally, we identify the spatial diffusion coefficient as $D = k_B Q/\Gamma$ and we are confronted with a specific form of the famous fluctuation–dissipation theorem revealing that damping always causes fluctuations or in other words noise with an intensity given by the temperature Q.

For chemical processes described by reaction–diffusion processes as referred in Eq. 1.3 such thermal noise can play a dominant role and can lead to nucleation, decay or even stabilization of chemical patterns [18, 95, 52]. Living creatures or structures in vivo like cells or neurons exist far beyond the thermal equilibrium and thus no fluctuation–dissipation theorem holds. However, noise is an important driving mechanism and life would not have been developed without it. In such biological systems the origin of noise can be deeply hidden in the intrinsic metabolistic interplay and surrounding temperature is only one of various factors that tunes noise intensity or correlations. For neurons noise is often considered as a discrete and random input sequence coming from the active neuronal environment similar to molecules hitting a bigger particle randomly [21].

In neuron models this kind of noise enters as a stochastic current in the activator equation. Another source of fluctuations is channel noise that can enter the equations for modeling the dynamics of ion channels through active membranes. A huge number of such channels transport ions in a not completely identical way. The statistics of opening and closing events has therefore a certain width which lead to a noise term in the corresponding Langevin equation [156]. A similar perturbation acting on a much slower time scale is the extracellular ion concentration that can force ion channels to release ions when it has reached a concentration threshold. The latter is studied in chapter 3.

In the following, whenever noise enters in considered models, we assume Gaussian and white noise. Thus the intensity distribution obeys a Gaussian distribution and the temporal distance between two stochastic events has no lower or upper limitation according to a power spectrum that includes the whole frequency range.

2.2 Escape from a parabolic potential

In this chapter we estimate the mean escape time from a stable state induced by noise. We will give a simplified approach coming from an reduced and abstract model to find the expression for the famous Kramers rate. The underlying neuron model is a Leaky–Integrate–and–Fire model (LIF) that work as follows. External perturbations affects the polarized membrane potential until a certain threshold is reached. Then, by definition, a spike event is produced and after some time (which can also be zero) the potential is reset to a value close or at the resting potential. This procedure can be described by an Ornstein–Uhlenbeck process [147] given by the equation

$$\dot{x} = -x + \sqrt{2}\,(t),\qquad(2.5)$$

for x interpreted as the membrane potential. This is exactly the same dynamics as given in Eq. 2.1 when x is interpreted as a velocity $\dot{x}$ and as the noise intensity. In fact, Ornstein and Uhlenbeck did not consider activity of a neuron but the velocity of a Brownian particle which is confined in a parabolic velocity potential due to the fluid's friction. However, for x being a spatial coordinate Eq. 2.5 describes the position of a particle moving in a spatial parabolic potential. When the potential has a more complicated structure, than this dynamics hold at least for the linearized region around a minimum. For Eq. 2.5 the potential reads $U(x) = x^2/2 + \text{offset}$ with a minimum at $x = 0$ that constitutes the only fixed point of the dynamics. We parameterize a parabolic potential that prevents free diffusion of the particle in a more generals form as

$$U(x) = x^2 + x + \text{offset} = \left((x + \frac{}{2})^2 - \frac{^2}{4\ 2}\right) + \text{offset}\,.\qquad(2.6)$$

The parameters and can be any real numbers. The only condition we put is > 0 in order to make sure that the potential possesses a minimum as the equilibrium state for a noise–free particle. The offset is set to be zero for further calculations, since it has no impact on the physical situation.

We set the domain of definition as $x \in (-\infty, b)$ and the initial position of the Brownian particle at $x_i = a$. An absorbing barrier is set at $x_{\text{abs}} = b$ which is by definition greater than a. If the position of the barrier is smaller than the minimum's position ($b < x_{\min}$) then the Brownian particle performs regular cycles containing deterministic motion to the barrier, crossing the barrier and re-infection at x_i. Noise disturbs the regularity of this procedure and plays therefore a destructive role.

2 Stochastic escape from a fixed point

For a barrier beyond the minimum ($b > x_{\min}$) a noise–free particle ends in the minimum of the potential resting there forever. Stochastic perturbations allows the particle to leave the minimum and escape over the barrier in a finite time. This situation is schematically depicted in Fig. 2.1. For an arbitrary potential the mean time to cross the barrier b the first time depending on the initial position a was calculated 1933 by Pontryagin [112]. In our notation it reads:

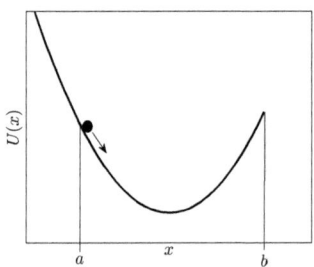

Figure 2.1: Scheme of the potential $U(x)$ with the initial state a and the absorbing barrier b.

$$T = \frac{1}{\varepsilon} \int_a^b dx\, e^{U(x)/\varepsilon} \int_{-\infty}^x dy\, e^{-U(y)/\varepsilon} , \qquad (2.7)$$

and is called the mean first passage time $T = T(a \to b)$ (MFPT). A simple approach to arrive at this expression is given in Appendix A. We apply now the parabolic potential (2.6) and calculate the inner integral first:

$$I = \int_{-\infty}^x dy\, e^{-U(y)/\varepsilon} = \int_{-\infty}^x dy\, \exp\left(-\frac{\alpha}{\varepsilon}(y + \frac{\beta}{2\alpha})^2\right). \qquad (2.8)$$

By substituting $z^2 = \frac{\alpha}{\varepsilon}(y + \frac{\beta}{2\alpha})^2$ it remains

$$I = \sqrt{\frac{\varepsilon}{\alpha}} \int_{-\infty}^{g(x)} dz\, e^{-z^2} , \quad \text{where } g(x) = \sqrt{\frac{\alpha}{\varepsilon}}\left(x + \frac{\beta}{2\alpha}\right). \qquad (2.9)$$

Using the error-function, defined by $\operatorname{erf}(x) = \frac{2}{\sqrt{\pi}} \int_0^x dz\, e^{-z^2}$, we obtain

$$I = \frac{1}{2}\sqrt{\frac{\pi\varepsilon}{\alpha}}(1 + \operatorname{erf}[g(x)]). \qquad (2.10)$$

2.2 Escape from a parabolic potential

The expression for the MFPT then reads

$$T = \frac{1}{2}\sqrt{\frac{\pi}{\sigma}} \int_a^b dx \exp\left(-(x+\frac{\sigma}{2})^2\right)\left(1+\mathrm{erf}[g(x)]\right). \qquad (2.11)$$

With the same substitution we have performed for the inner integral, we relegate any parameter dependency from the integrand

$$T = \frac{\sqrt{\pi}}{2}\int_{g(a)}^{g(b)} dz\, e^{z^2}\left(1+\mathrm{erf}[z]\right). \qquad (2.12)$$

For simplification we initiate the escape process in the minimum at $a = x_{\min} = -\frac{\sigma}{2}$. Typically, the escape times are given with respect to the barrier height. Here we have $\Delta U = U(b) - U(x_{\min}) = (b-x_{\min})^2$. Additionally we use a general expression for the curvature of the potential: $U''(x) = 2$ and thus we can write

$$T = \frac{\sqrt{\pi}}{U''}\int_0^{\sqrt{\Delta U/\sigma}} dz\, e^{z^2}\left(1+\mathrm{erf}[z]\right) \xrightarrow[\sigma \ll \Delta U]{} \frac{1}{U''}\sqrt{\frac{\pi\sigma}{\Delta U}}e^{\frac{\Delta U}{\sigma}}. \qquad (2.13)$$

The last approximation leads to the famous Kramers formula, calculated in 1940 [73]. In figure 2.2 both expressions are compared in double–logarithmic presentations. Solid lines correspond to the integral in Eq. 2.13, whereas dashed lines refer to the Kramers expression. It becomes clear that as soon as the barrier height and the noise intensity are of the same order of magnitude, Kramers approximation deviates from the exact integral expression.

However, for any $\sigma \lesssim 0.1\Delta U$ Kramers formula seems to be an excellent estimate. Note, that small noise and large barrier heights lead to extremely long escape times. Reminding noise as an microscopic representant for environmental temperature such long times explain the stability chemical compounds that do not decay due to thermal activation within time scales of biological processes for instance.

2 Stochastic escape from a fixed point

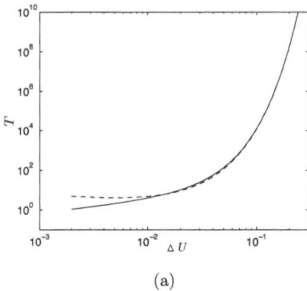 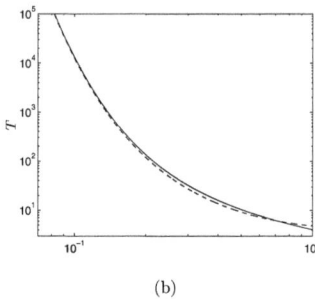

(a)   (b)

Figure 2.2: Scaling of escape times from Eq. 2.13; dashed lines refer to Kramers approximation. (a): versus potential height (for = 0.01), (b): versus noise intensity (for $\Delta U = 1$).

2.3 Mean escape time in the FHN dynamics

2.3.1 Probability density distribution in phase space

After the treatment of the escape from a parabolic potential we consider the more extensive situation pertaining the escape from a steady state in an excitable system. Without having an underlying potential it is not possible to find an analytical expression as given in Eq. 2.7 for such systems. However, in analogy, the fixed point can be consideres as a minimum in a potential landscape while the potential barrier which has to be overcome corresponds to the threshold around that fixed point. We consider a FHN system as a prototype model for an excitable dynamics in the form

$$\dot{u} = \frac{1}{\ }(3u - u^3 - v) + \sqrt{2\ }\ _u$$

$$\dot{v} = \ + 1 + u + \sqrt{2\ }\ _v .$$
(2.14)

Compared to the FHN system from Eqs. 1.2 these equations are written slightly different but describe the same type of behavior. The inhibitor dynamics is even simplified due to the lack of any dependency on v itself. Thus the inhibitor nullcline is perpendicular to the abscissa and crosses the cubic activator nullcline at a point controlled by the value of . It defines the location of the fixed point at ($u^* = - - 1, v^* = \ ^3 + 3\ ^2 - 2$). For any $\ > 0$ the system is excitable and thus we identify this parameter as the excitability. Noise enters in both variables with the same intensity . Again it is assumed to be white and Gaussian.

Each trajectory following the dynamics given in Eqs. 2.14 has its individual noise

2.3 Mean escape time in the FHN dynamics

realization and $u(t)$ and $v(t)$ are stochastic variables. Hence, it may be worthwhile to take a look at the distribution of an ensemble of trajectories, i.e. the density of probability distribution $p(u,y,t)$ to find a stochastic particle within an interval $(u+du, v+dv)$ at time t. The evolution of the probability density in space and time is given by a Fokker–Planck equation [123]. For the FHN system it has been investigated for oscillatory behavior [145] or for the stationary excitable regime [72]. The specific Fokker–Planck equation corresponding to the Eqs. 2.14 reads:

$$\frac{\partial}{\partial t} p(u,v,t) = \left(\frac{\partial^2}{\partial u^2} + \frac{\partial^2}{\partial v^2} \right) p(u,v,t)$$
$$- \frac{1}{\varepsilon} \frac{\partial}{\partial u} \left[\left(3u - u^3 - v \right) p(u,v,t) \right] - \frac{\partial}{\partial v} \left[(\ + 1 + u) p(u,v,t) \right], \quad (2.15)$$

where the upper line resembles the diffusion equation in two dimensions with the diffusion coefficient according to the stochastic force terms in the Langevin equations with its noise intensity. The second line contains the drift terms including the reaction parts of Eqs. 2.14. For boundary conditions that do not allow a leakage of probability (no–flux or periodic) a stationary solution exist with $\partial_t p^0(u,v) = 0$. It can be found by numerical methods using finite–differences or finite–element schemes. For some cases it can be computed much faster by running the Langevin equations directly for numerous trajectories, which populate the phase space due to noise by degrees. This has been done to obtain Fig. 2.3 (a) where the density information is coded as gray levels. Though this picture is a snap shot of current positions at a certain time it presents the stationary solution, yet. After a transient time, which is temporized for the calculation, a balanced fraction of particles in phase space has been developed forming a stationary distribution. As it can be seen in the figure the main part of probability accumulates around the stable cubic branches of the activator nullcline (adumbrate as dashed lines in the figure). Especially along the branch that ends into the stable fixed point the probability to find a particle is higher.

The distribution of probability in phase space provides information about the dynamical structure of the underlying system. In order to discover such dynamical features we consider at first the marginal probability density $p^0(v) = \int du\, p^0(u,v)$ shown in Fig. 2.3 (b) in the upper panel. Although the FHN model is a non–potential system we presume $p^0(v)$ to be connected to a corresponding one–dimensional pseudo–potential via $p^0(v) \propto \exp(-U_p(v)/\ )$, illustrated in Fig. 2.3 (b) lower panel.

The maximum of probability density is represented by a minimum in the pseudo–potential, which is not necessarily located at the original fixed point position due to the presence of noise. Depending on the surrounding force field probability may accumulate on a shifted position close to the fixed point [98]. Nevertheless, it seems likely apply the approach we have used in the previous section. In the following we discuss a potential based description that enables us to treat the escape from the

2 Stochastic escape from a fixed point

stable rest state of the FHN system as an escape process over a potential barrier.

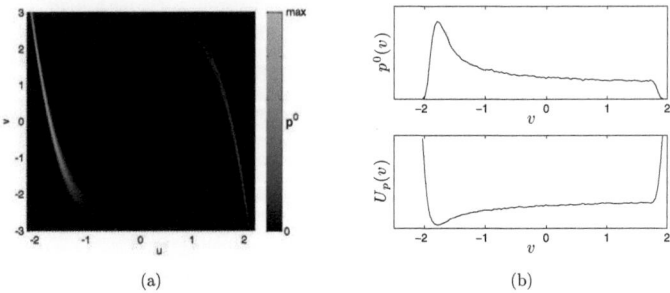

Figure 2.3: (a): Stationary probability distribution as gray levels according to Eq. 2.15 for = 0.1, = 0.1 and = 0.05. Nullclines are adumbrated as dashed lines. (Colors online) (b): marginal density for v (top) with its corresponding pseudo–potential (lower panel)

2.3.2 Complete time scale separation

It is reasonable to reduce the two–component dynamics to one dimension, in which an underlying potential always can be written. This method is due to Lindner et al. [87]. By adiabatic elimination (→ 0) Eqs. 2.14 can be reduced to the slow dynamics along the stable branches of the cubic nullcline given by

$$\dot{v} = (\ +1) + u_{l,r}(v) + \sqrt{2}\ \ v,$$

$$\text{with} \quad u_{l,r}(v) = \mp 2\cos\left(\frac{1}{3}\arccos\left(\pm\frac{v}{2}\right)\right). \tag{2.16}$$

By assuming transitions between the stable branches to be infinitely fast we need to treat two decoupled equations instead of the coupled two–component FHN system. The indices mark the left and right branch. The force terms on the right hand side of the inhibitor equation correspond to the potentials

$$U_{l,r}(v) = -(\ +1)v - \frac{3}{4}u_{l,r}(v)\bigl(v - u_{l,r}(v)\bigr). \tag{2.17}$$

Both potentials are qualitatively different. On the left branch the potential possesses a minimum retrieving the minimum from Fig. 2.3. In contrast, the right handed potential is ramp–like and describes the relatively slow course along the right cubic nullcline branch, where the system is in the activated state for a quasi–

2.3 Mean escape time in the FHN dynamics

deterministic finite time. We will consider only escape from the first potential, whose minimum, located at v^*, coincides with the stable fixed point of the excitable dynamics. For $\to 0$ a clearly defined boundary exists, that determines the height of the potential barrier.

We regard perturbations as instantaneous shifts of the nullclines in phase space. For the case of inhibitor noise the corresponding nullcline fluctuates around its deterministic value. If this linear nullcline crosses the Hopf–bifurcation point, located at the minimum of the cubic nullcline ($u = -1, v = -2$), the corresponding fixed point becomes unstable and thus repels trajectories. Hence, we consider the location of the cubic nullcline's minimum as the threshold position playing the role of the absorbing boundary.

The barrier of the potential that has to be overcome by noise is determined as: $\Delta U_l = U_l(v = -2) - U_l(v^*)$. Note, that ΔU_l depends only on . In the limit of low noise the mean escape time is given in [87] as

$$T = \frac{1}{U_l''(v^*)}\sqrt{\frac{}{\Delta U_l}}\, e^{\frac{\Delta U_l}{\sigma}}, \quad \text{for} \quad \ll \Delta U_l, \tag{2.18}$$

according to the expression in Eq. 2.13. Here, the curvature is derived at the minimum of the potential, where a parabolic shape can be assumed. In the mentioned publication also expressions beyond the Kramers range of validity are given, which we do not consider here.

Instead, we study the scaling of the parameters noise intensity   and the excitability   as shown in Fig. 2.4. Therein the solid lines refer to Eq. 2.18 and circles present numerical results. Additionally, dashed lines representing typical scaling are drawn. In Fig. 2.4 (a) where the escape time is plotted versus noise we chose a excitability value of $= 0.2$ (and $= 10^{-4}$), which leads to a potential barrier of $\Delta U_l = 0.0084$. For any noise intensity of the same order of magnitude or less ($\lesssim \Delta U_l$) the numerically found escape time matches the Kramers time. The leading dependency on   is given by the exponent $\propto \exp(\Delta U_l/\ )$, shown as dashed line.

In Fig. 2.4 (b) the dependency on   is displayed. Here, less numerical points are created. This is due to the fact that for ≥ 0.5 the potential barrier is more than ten times larger than the noise intensity leading to extremely long escape times that may be longer than typical phd positions. In the potential in Eq. 2.17 we have a leading order in the excitability of (^{4}) which produces a scaling of $T \propto \exp(c\ ^4)$ where c is an arbitrary fit parameter. This scaling is drawn in the figure as dashed line.

2.3.3 Spatially extended excitable and sub–excitable medium

As mentioned in section 1.3 the spatially extension of excitable systems creates an active medium in which a superthreshold perturbation leads to a local activation

2 Stochastic escape from a fixed point

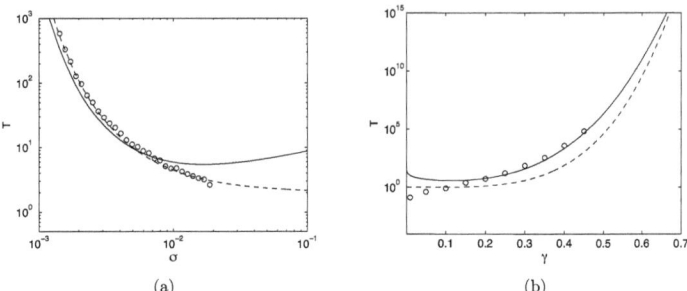

Figure 2.4: Mean escape times from the FHN fixed point. Solid lines are due to the result from Eq. 2.18 compared to numerical simulations (circles) and scaling behavior (dashed lines). Used parameters values in (a): $\gamma = 0.2$ and $\epsilon = 10^{-4}$ and in (b): $\sigma = 0.01$ and $\epsilon = 0.01$.

that can ignite the surrounding medium forming a spreading pattern. These nucleation processes strongly depend on the parameters of the medium such as diffusion coefficient or excitability [53]. The excitable regime in the local dynamics splits in three cases in the spatially extended situation. First the nonexcitable regime, where perturbations never trigger a superthreshold reaction in the medium. Here, the diffusion for example is so fast, that each local distortion is quickly balanced. In contrast in the excitable regime local perturbations grow easily and provide widespread patterns. In between these types of behavior there is a region of subexcitability, characterized by local super–threshold excitations that do not form extended patterns though. Also initial extended structures with a finite size decay soon.

There are various works referring to that problem also in two dimensions. In the photo–sensitive BZ reaction it was shown [75] that the excitability of the medium can be controlled by tuning the light intensity illuminating the chemical reaction. Considering moving wave segments in this reaction, one finds illumination intensities for specific wave sizes, which make those wave segments grow or shrink [125]. The higher the illumination level the larger the wave segment must be to not decay. There exists an illumination intensity corresponding to a critical excitability that separates the excitable and subexcitable region.The dependency between size and excitability saturates at a finite excitability value beyond which every two–dimensional structure shrinks and finally disappears. Here, the system is definitely in the subexcitable regime for two–dimensional patterns.

Showalter et al. invented a technique for stabilizing wave segments of a certain size in the lightsensitive BZ reaction [95]. It is based on a mechanism of feeding back

2.3 Mean escape time in the FHN dynamics

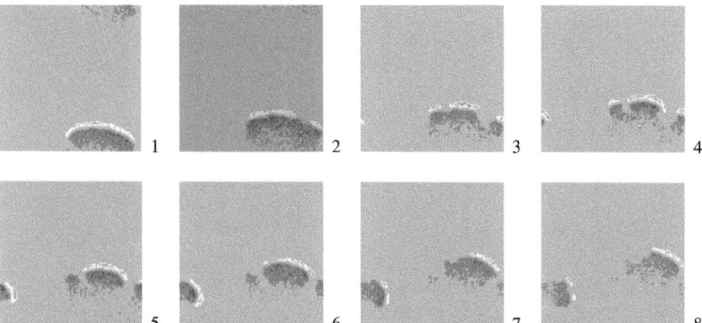

Figure 2.5: Snapshots of stabilized moving wave segments due to simulation of a spatial extended FHN system. Stochastic fluctuations lead to a wave break–up. The products may grow forming a new stable wave segment or shrink until they disappear. (Colors online)

the current wave size as an illumination information. We reproduced this technique for simulations of the spatially extended FHN system where activator noise is added. Some snapshots of a propagating stabilized wave segment are shown in Fig. 2.5. In the presented situation the presence of noise leads to a break–up of the segments. However, since the stabilization mechanism keeps the total wave size fix, some of the pieces survives and grow at the cost of remaining pieces which then decay.

Increasing the light intensity or the excitability, respectively, leads to a system state where not even one–dimensional patterns can move nor survive. This is illustrated in Fig. 2.6, where under the influence of noise activation events occur and either decay without creating connected patterns in the subexcitable case (Fig. 2.6 a) or form waves in the excitable case (Fig. 2.6 b). In the FHN system as written in Eqs. 2.16 is the excitability parameter and thus plays the role of light intensity. The smaller is the better waves can propagate.

In the following we give a brief estimation for the transition parameters from subexcitable to excitable behavior. We consider N FHN elements coupled to their neighboring units arranged along a ring so that so asymmetry at edges occur. For the case of activator coupling the deterministic version of Eqs. 2.14 is extended to

$$\dot{u}_n = \frac{1}{\ }(3u_n - u_n^3 - v_n) + \ (u_{n+1} + u_{n-1} - 2u_n)$$
$$\dot{v}_n = (\ + 1) + u_n, \quad \text{with} \quad n = 1, 2, ...N. \qquad (2.19)$$

For the moment, we set the coupling to one, so that the only parameters are and

2 Stochastic escape from a fixed point

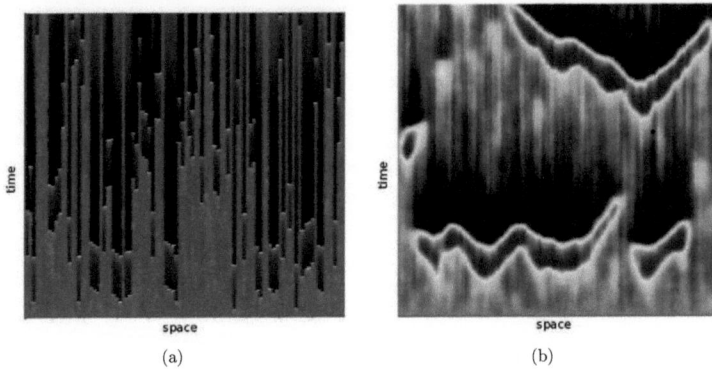

Figure 2.6: Space–time plots of coupled noisy FitzHugh–Nagumo elements. (a): subexcitable regime, (b): excitable regime. (Colors online)

. In numerical simulations for $0 < \varepsilon < \varepsilon_{\text{crit}}$ running waves can be found after an initial superthreshold perturbation, corresponding to the situation shown in Fig. 2.6 (b) while for $\varepsilon > \varepsilon_{\text{crit}}$ an activated local FHN element will not excite their neighbors as shown in Fig. 2.6 (a).

We interpret the diffusive coupling as a signal coming from an activated neighboring element so that for instance the $(n+1)$th element is activated and the $(n-1)$th element remains in the fixed point. The maximal output an activated element produces, is the difference $u_{\text{max}} - u^*$. For a trajectory moving straight from the nullcline's minimum, which was assumed to be the absorbing boundary and which is close to the fixed point, arriving at the right stable branch, we find for $u_{\text{max}} = 2$. Therewith the nullcline of the n-th unit is shifted to $v_n = 3u_n - u_n^3 + (u_{\text{max}} - u_n)$ illustrated in Fig. 2.7 (a) where a sector of the phase space near the fixed point is displayed. The thin dashed line corresponds to the unperturbed cubic nullcline and the thin solid line shows the shifted nullcline. The perpendicular lines demonstrate two values for the excitability that lead to two different fixed point locations. Relative to them, the shift of the cubic nullcline causes an escape of trajectories from the fixed point, as shown for the lower trajectory (thick solid line in the figure) or to a direct decay to the new fixed point as exemplified by the upper trajectory in the figure.

As an upper limit for the excitability beyond which the system is definitely subexcitable, we estimate parameter values that shift the minimum of the nullcline exactly to the inhibitor position of the former fixed point $v_{\text{min}}^p = v^*$. The unperturbed inhibitor fixed point value we already estimated as $v^* = \varepsilon^3 + 3\varepsilon^2 - 2$ and for the

2.3 Mean escape time in the FHN dynamics

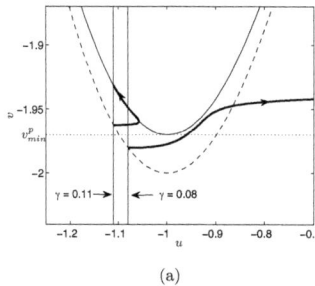 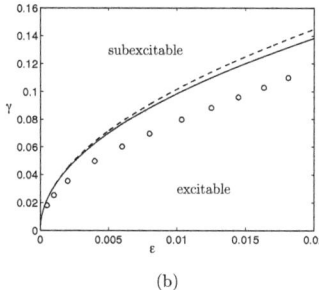

Figure 2.7: (a): Shift of the cubic nullcline (thin solid line) from its former position (dashed line) when a neighboring element is excited. Trajectories are drawn (thick solid lines) for two values of . (b): Line of critical excitability from Eq. 2.20 with its expansions to the linear order (dashed) separating the parameter space in the different regions of excitability. Circles represent numerical results.

shifted minimum we find $v_{min}^p = u_{max} - 2/(3\sqrt{3})(3-\ )^{3/2}$. Solving for yields

$$_{\text{limit}} = 2\cos\left(\frac{1}{3}\arccos\left[\frac{v_{min}^p(\ )}{2}\right]\right) - 1 = \sqrt{\ } - \frac{1}{6} + \ (\ ^{\frac{3}{2}}). \qquad (2.20)$$

This relation separates roughly the parameter space of and in the mentioned regions of subexcitablity (> $_{\text{limit}}$) and excitability (< $_{\text{limit}}$). For a finite time scale separation the vector field belonging to Eqs. 2.19 gets a non-negligible component in v−direction. Thus trajectories can directly return to the new fixed point even when the minimum is shifted beyond the fixed point position. The expression given in Eq. 2.20 is therefore valid asymptotically for → 0 and the critical value, that separates the two regimes of excitability can be estimated as $_{\text{crit}}$ < $_{\text{limit}}$ if > 0. Fig. 2.7 (b) shows the − parameter plane including the relation of Eq. 2.20 which is compared to numerical results, shown as circles. As expected, the smaller the better the critical line from Eq. 2.20 matches the simulations.

Restating the approach from the local FHN element activated by noise we consider noise induced excitations in FHN units coupled to a passive environment from the same type. Anew we are interested in the escape times of such coupled excitable elements that either lead to running activation waves in the spatially extended version or to single independent activation events in the subexcitable regime. The following

27

2 Stochastic escape from a fixed point

estimation neglects all kinds of successive dynamics triggered by a very first escape event but exclusively considers a single element that gets activated. This element might be interpreted as infinitesimal small compartment of a continuous diffusively connected medium. It can be also interpreted as a participant of a network that consists of discretely coupled neighbors as formulated for a chain in Eqs. 2.19. For the latter situation we assume that noise affects only the considered unit while their two neighbors remain completely unaffected by any distortion.

Writing the inhibitor equation by eliminating adiabatically the fast dynamics ($\to$ 0) as done in Eq. 2.16 and including the discrete coupling from Eqs. 2.19 with neighboring units located in the fixed point $u_{n\pm 1} = u^*$ we obtain a inhibitor potential which is biased due to the coupling . It reads:

$$U(v) = -(+ 1)v + \frac{3}{4}u(v)\left((1 - \frac{2}{3})u(v) - v + 2u^*\right), \qquad (2.21)$$

with $u(v) = r\cos\left(\frac{1}{3}\arccos\left[\frac{4}{r^3}(2u^* - v)\right]\right)$ and $r = -2\sqrt{1 - \frac{2}{3}}$.

Following the method that we presented in the previous section we consider the

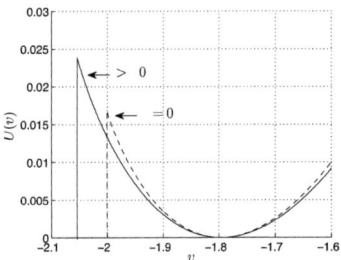

Figure 2.8: Potential according to Eq. 2.21 for = 0 (dashed line) and > 0 (solid line).

minimum of the nullcline $v_{\min}$ as absorbing boundary and the fixed point value v^* as the initial location. Note, that the latter does not depend on the coupling strength while any > 0 increases the absorbing boundary and thus the potential barrier height (see Fig. 2.8). The increasing barrier height with higher is intuitively clear, when we remind, that the neighbors rest in the passive and non–activatable state to which the dynamical unit is coupled.

The escape times can be calculated using Kramers formula (Eq. 2.18) and applying the potential from Eq. 2.21. The outcome is plotted versus the excitability in Fig. 2.9 (a) for a coupled unit (solid line) compared to the uncoupled case (dashed line). As estimated the escape time for a single unit coupled to inactive neighbors is larger than for a single unit. That remains true also in the small range where the Kramers prediction fails. For larger excitability values the difference is considerable and can exceed one order of magnitude. For ≤ 0.1 the potential barrier is comparable to the noise value and numerical results deviates from the Kramers curve.

In Fig.2.9 (b) the dependency on the coupling scaled with is shown for two noise intensities. The upper (solid) curve corresponds to a low noise value while for the lower (dashed) curve is three times larger noise intensity is chosen. Kramers approximations is in a good agreement with numerical results except for too small at the curve with higher noise strength. It can be seen, that the coupling antagonize higher noise values. The difference in the escape times even increases with higher coupling which indicates a non–linear dependency on the latter.

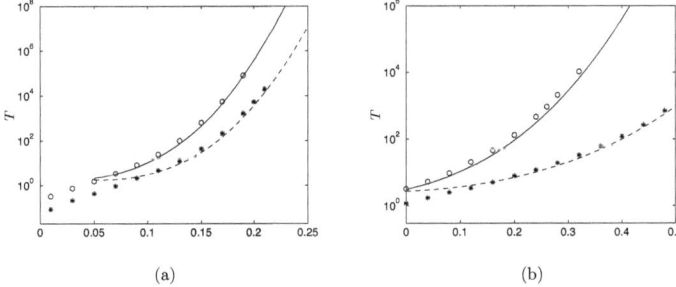

Figure 2.9: Mean escape times of a FHN unit coupled to passive neighbors calculated with Kramers formula using the potential from Eq. 2.21 (solid and dashed lines). Symbols stand for numerical results. (a): coupled and uncoupled case is compared, solid line and circles: = 0.1, dashed line and stars: = 0, both: = = 1 10^{-3}. (b): two noise intensities are compared, solid line and circles: = 1 10^{-3}, dashed line and stars: = 3 10^{-3}, both: = 0.1 and = 1 10^{-3}.

2.4 Conclusions

The chapter shall give an overview how noise acts on excitable systems and how it is to quantify. However, the presented concepts are nothing new. Due to the general relevance of such processes they are often treated in literature as the reader may have seen in the references. The method used in the whole chapter deals with an

2 Stochastic escape from a fixed point

approximated underlying potential structure and its relation to the nullcline picture in phase space. We expand this method to a coupled two–component excitable system.

Excitable dynamics are characterized by a rest state that can be left due to the influence of noise. At the beginning we studied the escape from a parabolic potential with an absorbing barrier being the simplest model for excitable behavior. For this case we gave an estimation to arrive at the famous Kramers formula to quantify the escape over potential barriers. We expand this approach for the FitzHugh–Nagumo dynamics. Under the condition that the activator and inhibitor time scales are well separated an underlying potential can be constructed allowing us to use Kramers expression. We calculate the escape time and find scaling behavior for the excitability parameter.

For a spatial extended excitable medium a differentiation of the character of excitability is needed in regard to patterns formation. In the subexcitable regime no extended patterns can develop and local perturbations decay. Stable and extended structures are only possible when the excitability parameter is below a critical value. We derived an approximate formula describing the transition line between these two regimes in the parameter plane. Finally, we described the excitation of an element which is surrounded by passive medium. The passive neighborhood enters in the potential which was written for the local unit. Higher coupling increases the potential barrier and lead to longer escape times.

3 Dynamical structures in a heterogeneous active medium

3.1 Introduction

In the coming chapter we will successively expand the model which we have studied in the last section of the previous chapter. In pure mathematical terms a third dynamical equation will be affiliated to the two–component excitable stochastic FHN dynamics, introduced in section 1.2. The three dimensional phase space of the extended model will be investigated and its most important bifurcations will be presented. Furthermore, this system will be used to arrange a spatial extended heterogeneous medium in which pattern formation will be discussed.

Although we will keep the level of abstraction high, the model is based on biological mechanism which is related to the original meaning of excitable systems that is the neuronal activity and ion channel dynamics within cells. As we introduced in section 1.2, electrical activity of a cell is represented by depolarization of its membrane potential which is well described by the mentioned HH. But also simplified model systems using a smaller amount of variables and control parameters [67, 60], are able to specify the many aspects of neuronal dynamics. But an inevitable assumption and simplification has been made for these prototypical neuron models in terms of ionic currents. For example, it was assumed that in spite of transmembrane currents both, extracellular and intracellular ionic concentrations remain unchanged during depolarizations. Such a simplification is natural and acceptable if individual neurons or segments of an excitable medium during sufficiently short activation events are considered.

However, in other cases it will be not realistic. For example, there is the experimental evidence that extracellular concentration of potassium ions can rise significantly during intensive neuronal activity [139, 20]. Detailed models include severals sources and sinks for extracellular potassium ions. The governing currents are the ion–flow through the neuronal membrane, lateral diffusion and the uptake by the surrounding glia network [17, 148]. Beside supporting the neurons glia cells have different duties such as spatial stabilizing of the neuronal network or isolating their electrical potentials. Therefore they have similar properties like neurons such as a resting potential or the response to neuronal firing events which may enhance or surpress the neuron's activity [74]. So, they can activate potassium pumping when the ion level rises considerably [22]. Models that take this positive feedback into account show a strong increase of extracellular potassium concentration up to 80 mM [164, 161, 51]. It is speculated that this excessive elevation of potassium concentration potentially is an important element of the mechanism of epileptic seizure development [9, 108].

In neuro–physiology there is a lot discussion about the functionality of extracellular ions. It is shown that the decomposition of potassium is not only due to diffusion. Different pathways must be involved in this process [154, 39, 105, 24]. More recent models addressed the detailed neuronal morphology [64, 65] or the role of specific ion channels in formation of self–sustained bursting behavior that may even lead to pathological neural activation [9, 37]. The wider list of modeling issues on this topic was recently reviewed in [36].

The aim of the current chapter is to construct a low–dimensional model exhibiting transparent dynamical regimes and possessing an adequate number of control parameters. The exact biological situation, however, would require a rather complex model which might include numerous dynamical equations. The effects of potassium mediated coupling were investigated using the Hodgkin–Huxley type model [115, 116]. It was shown that such rather simplified, but still quantitative model reproduces the main features of small ensembles of potassium–driven neurons. With the approach presented here, we want to dispense with the possibility of a quantitative description. In order to discover basic mechanisms and the dynamical rules of larger networks we will use a well understood model and study the mentioned biological situation under aspects of excitable, oscillatory and bistable behavior.

Therefor, we construct such a model in the form of an extended FHN system with an additional equation describing the dynamics of extracellular potassium. Since our model inherits the key features of a FHN element, it is physically transparent and tractable. It thus provides the better chance to understand primary underlying nonlinear mechanisms governing the local activity as well as the formation of spatio–temporal patterns in large networks. Furthermore, compared to leaky integrate and fire models (see section 2.2) the FHN unit includes the whole reset mechanism of the neuron and is therefore appropriate to represent the essence of single neuronal spike events. Some dynamical behavior which we will observe later on in this chapter refers to the excited state of the FHN which is not provided by the LIF (see section 2.2) model [16, 143].

Applied to interacting neurons we assume that the interaction of neurons is restricted to the chemical pathway. Coupling takes place indirectly due to the potassium concentration outside the cells, exclusively. In the extended model no explicit metric will be applied and thus no distance will be defined. Nevertheless, the neurons are assumed to be strictly separated and a direct contact of the action potentials is excluded. The biological fundament for this assumption yields glia cells surrounding the neurons. The medium consists of neurons and extracellular space which contains such glia cells for instance. Thus it is heterogenous and signal processing or pattern propagation is slower than in a homogeneous excitable medium.

3.2 Abstract model for a potassium–driven neuron

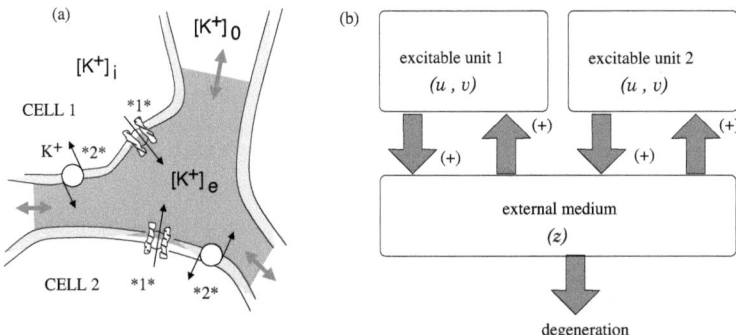

Figure 3.1: (a) Schematic representation of the potassium interchange between closely located cells and extracellular space, (b) structure of the functional model with the corresponding variables.

The patterns appearing in the spatial extended system show phenomena, which are reminiscent of chemical experiments in which comparable heterogeneous situations like two–layer systems or chemical oscillators moving in a diffusive environment have been studied [10, 141] and which we already referred in section 1.3. Also some of the presented structures are comparable to patterns reported in studies on calcium ion release across the endoplasmic reticulum [28, 132]. In these systems clusters with a finite number of ion channels on the reticulum are related to single excitable elements in our model. After releasing the ions they diffuse freely in the cytoplasm forming different kinds of calcium waves.

In the last section of this chapter we attempt to classify observed spatio–temporal patterns according to the relation between the governing control parameters of the extracellular medium. Peculiar, patterns that cannot appear in a homogeneous FHN medium are randomly–walking spots, long living meandering excitations, anti–phase firing patterns, inverted spirals and waves.

3.2 Abstract model for a potassium–driven neuron

3.2.1 Biological background

We consider an environment which is schematically depicted in Fig. 3.1. We assume that there is a certain volume between the cells from which the ionic exchange with the outer bath is rate limited. For simplicity we assume that this volume is homogeneous and we denote the extracellular potassium concentration as $[K^+]_e$

3 Dynamical structures in a heterogeneous active medium

here. Intracellular potassium concentration is symbolized as $[K^+]_i$.

With time, particularly during firing events in neurons, the potassium channels, labeled as *1* in Fig. 3.1 (a), open and outward currents from the cells deliver potassium to the extracellular space. Thus, $[K^+]_e$ rises while the intracellular concentration $[K^+]_i$ decreases just slightly, because $[K^+]_i \gg [K^+]_e$. In the following we neglect the associated intracellular changes of the potassium concentration and assume that this concentration remains constant.

Beside this source of potassium ion corresponding to an inflow in the extracellular space, several uptake mechanisms are counteracting as mentioned in the introduction. First, there are $Na - K$ ATP pumps, labeled as *1* in Fig. 3.1 (a), which drives K^+ back into the cells in order to balance the potassium concentration to its equilibrium value $[K^+]_0$. Additionally, surrounding glia cells absorb potassium ions which will be subsumed to the previous process as a common decay term. The exchange of K^+ ions with the environmental extracellular bath is assumed to take place by a diffusion process. Hence, it is governed by the concentration difference between the exterior close to a potassium delivering cell and the bath. Then the balance of potassium concentration in the extracellular space can be described as follows:

$$W\frac{d[K^+]_e}{dt} = \frac{1}{F}\sum_{i=1}^{N}(I_{i,K} - I_{i,\text{deg}}) + ([K^+]_0 - [K^+]_e), \qquad (3.1)$$

where W is the extracellular volume per unit area of the membrane and N is the total number of cells being neighbors to this volume. The currents $I_{i,K}$ and $I_{i,\text{deg}}$ describe the influx and the degeneration of potassium ions due to pumps, channels and glia cells for the ith-cell. They are divided by Faraday's constant F. The second term $([K^+]_0 - [K^+]_e)$ models the effective diffusion of potassium to and from the bath. This balance equation (3.1) provides the basis for the qualitative description in terms of a functional model which we will introduce below.

Note, that the variation of the ratio between the extra- and intracellular potassium concentrations affects the corresponding (Nernst-) potential and, hence, the firing activity. Increase of extracellular potassium concentration may depolarizes the cell beyond the threshold and can evoke spontaneous firing. However, too high extracellular potassium concentration acts as a toxin and blocks the cell activity completely.

3.2.2 Local model

In the following we propose an abstract model that aims to mimic the qualitative reproduction of main effects arising if a variable extracellular potassium concentration is taken into account. The structure of the model according to the biological framework is schematically depicted in Fig. 3.1 (b). Excitable units without direct

3.2 Abstract model for a potassium–driven neuron

connections representing a number of neurons that contribute to the extracellular potassium increase. The labeled "external medium", in which the released potassium accumulates, stands for the extracellular space that separates the excitable cells. The dynamics of the external concentration have to fulfill the following conditions: The concentration level

(i) rises when one of the neighboring excitable units is in the active or firing state,

(ii) provides an additional stimulus to surrounding excitable units and

(iii) relaxes to an equilibrium state when no activation is received.

Let us first confine to a single neuron interacting with the external medium. Particularly, we will implement the activity of an excitable neuron by a FitzHugh–Nagumo system, written in the form:

$$\dot{u} = u - \frac{1}{3}u^3 - v \qquad (3.2)$$

$$\tau(u)\dot{v} = u + a - Cz, \qquad (3.3)$$

where ε controls the time scale separation of the fast activator variable u and the slow inhibitor variable v, as already introduced in the previous section. The operating regime of the FHN neuron is defined by a, playing the role similar to the applied current in ionic currents–based neuron models and is the excitability parameter (see section 2.3). External stimuli are due to synaptic excitatory or inhibitory coupling that works on a much faster time scale than ionic exchange with to extracellular environment. Therefore we model the synaptic input as noise which makes the parameter a fluctuate around its mean value a_0:

$$a = a_0 + \sqrt{2D}\,\xi(t), \qquad (3.4)$$

where $\xi(t)$ is white Gaussian noise with zero mean and intensity D.

As an extension we introduce an additional time scale $\tau(u)$ in (3.3) in order to control the two time scales, associated with firing (high level of u variable) and refractory state (low level of u) independently. Specifically, we apply the sigmoidal function

$$\Psi(u) = \frac{1}{2}\left(1 + \tanh\left(\frac{u}{u_s}\right)\right), \qquad (3.5)$$

which is sensitive to the current value u. It tends to zero for $u \ll 0$, and to one if $u \gg 0$, while u_s scales the transition between these states. For $u_s \to 0$ the sigmoidal function becomes a Heaviside function that distinguishes step–like between the excited and the resting states of the FHN neuron. With Eq. (3.5) $\tau(u)$ shapes as

$$\tau(u) = \tau_l + (\tau_r - \tau_l)\Psi(u) \qquad (3.6)$$

and takes the values $_l$ in the rest state and $_r$ in the excited state, respectively, for a u_s to be chosen as small.

Similar to the system given in Eqs. 2.14 the location of the cross section of the activator and inhibitor nullcline is exclusively controlled by excitability parameter a, as long as we consider the two–component system given by Eqs. (3.2–3.3) with $C = 0$. As a third component we add an equation to model the time evolution of the variable dimensionless extracellular potassium concentration, labeled as z, here. In the first instance, we do not consider any spatial dynamics and we write for the time evolution in accordance with Eq. (3.1)

$$\dot{z} = g(u) - z. \qquad (3.7)$$

Two terms govern the dynamics. The first one describes the production term with the release rate ≥ 0 that stands for the overall ionic currents outward the cells. These currents transport ions into the exterior when the unit is excited and channels are open, corresponding to $u > 0$. No ions are delivered when the channels are closed and the excitable unit is in the rest state or in the recovery period, corresponding to $u < 0$. Likewise for the time scales we use the function Eq. (3.5) as a trigger in Eq. (3.7) ($g(u) = \Psi(u)$). It set the production term close to zero if the unit is inactive or it enables the production for an active unit. The second term describes the ion loss by the mentioned processes like desorption through the glia network with a decay rate . The rest state corresponding to the steady state concentration $[K^+]_0$ is at $z^0 = 0$ and thus the concentration is always greater than zero ($z(t) \geq 0$).

The value of z enters in Eq. (3.3) with a factor C. It represents the depolarizing effect of an increased extracellular concentration. For a given non-vanishing value of z, this results in an effective decrease of the control parameter $a_e = a_0 - Cz \leq a_0$, which shifts the v-nullcline to higher values of u closer to the oscillatory behavior or even further.

The set of equations described above is dimensionless and, therefore, the relationship to the real biological situation can be only qualitative. However, for the sake of simplicity and to keep the connection with the original problem, we will use the terminology of neurophysiology further on in order to describe the dynamical behavior of the model as well as the meaning of control parameters. In the following we refer to system (3.2)-(3.7) as the FNK model that stands for the combination FitzHugh–Nagumo plus $[K^+]$.

3.2.3 Fixed points and bifurcations

In this section we want to study the main features of the local model in terms of the steady states and their stability. For $C = 0$ the FNK model converges to the original FHN system with a cubic and a linear nullcline generating one single equilibrium point at ($u^0 = -a, v^0 = a(a^2/3 - 1)$), which is stable for $|a_0| > 1$ and

3.2 Abstract model for a potassium–driven neuron

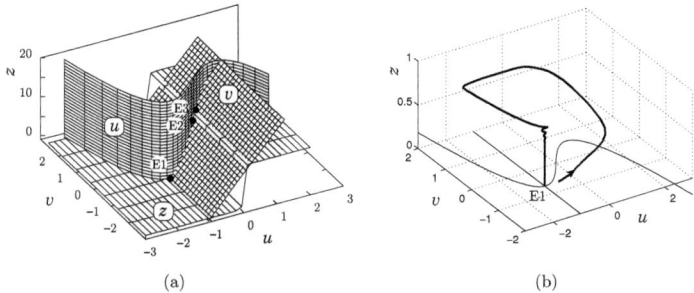

Figure 3.2: Three–dimensional phase space of the FNK model,
(a) nullcline surfaces labelled with their variables whose time derivative vanishes on them. The intersections of the cubic u-nullcline, linear v-nullcline and sigmoidal–shaped z-nullcline may provide one or three equilibrium points (E1, E2, E3).
(b) Representative trajectory near the stable equilibrium point E1 (thick line). The nullcline projections on the (u, v) surface are shown as thin lines.

unstable otherwise. The whole three–component FNK model, defined by Eqs. (3.2–3.7) possesses three nullcline surfaces, depicted in Fig. 3.2 (a). There they are labeled as u, v, and z according to the respective temporal derivative that vanishes on them. While the position of the cubic surface remains unchanged under the influence of the available parameters, the surface belonging to the step function can be manipulated by tuning , or u_s. When in the following numerical results concerning the local dynamics are presented then they are obtained by chosen parameter range of = 1.0...12.0, = 0.05...0.5 and u_s = 0.2.

The slope of the linear surface is controlled by C and its position by a_0, that is fixed to $a_0 = 1.04$ in the following to ensure to be in the excitable regime in the limit of $C \to 0$. The time scale parameters do not influence the nullcline surfaces and are set as = 0.04, $_l$ = (1), $_r$ = 1.0.

The fixed points are given by $\dot{u} = \dot{v} = \dot{z} = 0$ which yields

$$u^0 + a_0 = C \frac{}{2}\left(1 + \tanh\left(\frac{u^0}{u_s}\right)\right). \tag{3.8}$$

This expression has one or three real solutions for u^0. For a large C–range one fixed point remains close to $u_1^0 \approx -a_0$ at low z-level as a shadow of the steady state of the

37

3 Dynamical structures in a heterogeneous active medium

FHN system. Using a small u_s the right hand side of (3.8) becomes nearly a step function. Then the upper fixed point location can be given as $u_3^0 = C/ -a_0$ and the corresponding z-value is the highest z-level the system can reach ($z_3^0 = z_{max} = /$).

Taking C as the control parameter regulating the coupling strength to the exterior variable z this upper fixed point bifurcates at a critical value C_{crit} resulting from

$$C_{crit}\left(1 + \sqrt{1 - \frac{u_s}{C_{crit}}}\right) - u_s \text{atanh}\left(\sqrt{1 - \frac{u_s}{C_{crit}}}\right) - a_0 = 0, \qquad (3.9)$$

where $= \frac{}{2}$. That expression contains also a solution for a high C_{crit} where the lower fixed point annihilates in a saddle–node bifurcation. Thus, for any coupling higher than that C_{crit} the upper stable fixed point remains as the only attractor in the system. For a very high C the system always produce so much z by keeping the excitable unit active that it freezes in a completley depolarized state with a constant high level of $z = z_{max}$. For C values where three fixed points exist the unstable

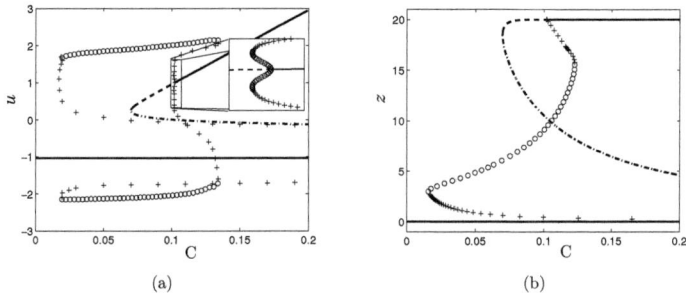

Figure 3.3: Bifurcation diagram for the coupling to the exterior C.
(a) lines mark the u values of fixed points – dashed: saddle, dashed–dotted: unstable, solid: stable; symbols show the extremal u elongation of limit cycles – circles: stable, crosses: unstable. The inset shows a zoom of the supercritical Hopf–bifurcation.
(b) z position of fixed points and limit cycles, lines and symbols as in (a).

equilibrium in between is located at $u_2 \approx u_s(2\ )/(C - 2u_s)$ which is only weakly dependent on C for our choice of the parameter range. In Fig. 3.2 (a) the positions of these three fixed points are illustrated and labeled with E1, E2 and E3.

In Fig. 3.2 (b) a representative trajectory for a very small C starting in the E1 vicinity is shown. The $\dot{u} = 0$ and $\dot{v} = 0$ nullclines are shown as projections on the (u, v) plane. The trajectory starts at a point beyond the basin of attraction of E1

3.2 Abstract model for a potassium–driven neuron

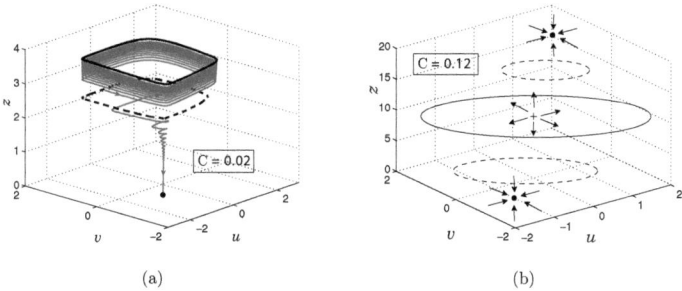

Figure 3.4: Phase space scenarios for two values of C.
(a) $C = 0.02$ – black solid: limit cycle, black dashed: unstable limit cycle, gray: two sample trajectories.
(b) $C = 0.12$ – schematic picture of attracting and repelling structures.

and therefore moves quickly to the cubic branch at positive u–values corresponding to the activated phase of the unit. During this time z increases and the trajectory also rises in vertical z–direction controlled by . After the typical activation course the trajectory comes back to the polarized state but it is still on a higher z–level. From this level, the phase point moves downward with the decay rate performing damped oscillations.

The mono– and multistable regimes in the FNK model are illustrated in the bifurcation diagrams shown in Fig. 3.3. Both diagrams show the same C parameter range whereas at the Y– axis the variables u and z are displayed. By linearization around the fixed points the number of (un–) stable directions can be determined. That is coded in the different line styles: Solid lines stand for complete stable fixed points (three negative eigenvalues), dashed–dotted lines correspond to complete unstable fixed points (three positive eigenvalues). For a small C–region there are fixed points which have two unstable directions with eigenvectors laying approximately in the (u, v) plane and one stable direction pointing approximately in z–direction. They are drawn as dashed lines.

As already mentioned the lowest fixed point is a relict from the FHN system and is stable until $C \approx 60$ (for $= 1$, $= 0.05$ and other parameter as mentioned above). This is far beyond the C–range we want to consider here. However, for $C = C_{\text{crit}} \approx 0.07$ two additional fixed points appear via saddle–node bifurcation close to the maximal z–level at $z_{\text{max}} = 20$. The upper point has one stable direction at first and bifurcates at $C \approx 0.1$ into a complete stable fixed point while demerging a limit cycle due to a supercritical Hopf–bifurcation.

39

Such limit cycles are also illustrated with their extremal elongation in the bifurcation diagram in Fig. 3.3 (a). They are symbolized with circles in case they are stable and with crosses when they are unstable. Note, that in the C, z–diagram the minimal and maximal elongation of the periodic orbits are nearly at the same z–level and thus only one line of symbols appears in the figure. The zoomed region of the Hopf–bifurcation in Fig. 3.3 (a) discloses the supercritical type. For a small region the created limit cycle is stable entraining two stable directions from the fixed point.

In the range of $C \approx 0.02...013$ there is a stable periodic orbit that annihilates with an unstable orbit on both sides of that range due to a saddle–node bifurcation. These unstable orbits merge with the low z–value and the high z–value fixed point branches after a second saddle–node bifurcation is passed, shown in the zoom region in Fig. 3.3 (a) for the upper branch. Thus the arriving orbits are stable and bifurcate via a supercritical Hopf–bifurcation. Within the C range containing the extended stable orbit the system is either bistable in which the lower fixed point and the limit cycle are regions in phase space where trajectories run into, or even three attractors exist; the lower and upper stable fixed point and additionally, the stable limit cycle in between. These cases are schematically depicted in Fig. 3.4.

The first case is presented in Fig. 3.4 (a) where a section of the phase space is shown with representing trajectories (gray lines) that are repelled from the unstable limit cycle (dashed black line). Depending on their initial conditions they either run asymptotically into the stable limit cycle (solid black line) or decay into the lower fixed point. For the sake of clearness in Fig. 3.4 (b) only the fixed points with their (un–)stable directions (arrows) and limit cycles are sketched (solid: stable limit cycle, dashed: unstable limit cycle). Essentially, it can be supposed that the unstable limit cycles separate the basins of attraction.

Beyond the limit cycle in a parameter range, where only the two stable fixed points exist, u and v also start to oscillate as a long living transient when perturbing the lower fixed point initiation as depicted in Fig. 3.5. In that case the z–level increases successively, lifting the system up to the maximal value. Thus the depolarizing spikes transform to polarization spikes, elongating from the depolarized state down to the former resting polarized level. The inverse firing process stops at a certain z level and reaches the stable steady state $E3$ corresponding to completely polarized neurons which are embedded in extracellular space contaminated by potassium.

3.3 Local dynamics under the influence of noise

Close to the polarized steady state trajectories spiral in the associated lowest fixed point as it can be seen in the example shown in Fig. 3.2 (b). That gives evidence of a non–vanishing imaginary part of the eigenvalues. The closer this fixed point is located at the Hopf–bifurcation value ($a_{\text{Hopf}} = 1$), the larger is the imaginary part. The effective threshold parameter is lowered by a z level greater than zero, provided

3.3 Local dynamics under the influence of noise

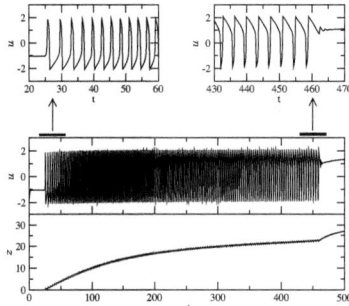

Figure 3.5: Transition to the upper stable state at $C = 0.2$ after the system is perturbed from the lower fixed point. During oscillations of u and v the value of z rises and drives the system to the upper fixed point. In insets (top row) the corresponding change of the spikes for u is illustrated.

by a larger value of C. That contributes to an increase of the imaginary parts of the eigenvalues. In Fig. 3.6 (a) the evolution in time of the three model variables is shown after a superthreshold perturbation from the lower stable fixed point for $C = 0$ (solid line) and $C > 0$ (dashed line). Fig. 3.6 (b) shows the zoomed region of the v–evolution in time after a spiking event close to the fixed point. It can be seen that subthreshold oscillations are much more pronounced for $C > 0$.

During the maxima of these oscillations the distance to values in phase space where a new excitation starts is reduced. Hence, during the moments of maximal elongation a weaker external forcing is sufficient to excite the next spike. This feature is important to understand how noisy input acts in this model. After the refractory time of a spike production there are few moments of larger probability for the next noise–induced firing. It resembles the behavior of so called resonate-and-fire neurons [59, 153] with subthreshold oscillations. But, in contrast, it occurs only after a spiking event, when the z–level has had time to rise. If at this state current noise values are too small to overcome the reduced threshold, then the next spike will occur after considerably longer time interval which can lead to bursting behavior [153, 129, 77].

The described subthreshold oscillations become more pronounced when the fixed point location gets closer to the Hopf–bifurcation and the system can be excited easier. The spectral power density $S(\ )$, that measures how much power is distributed

41

3 Dynamical structures in a heterogeneous active medium

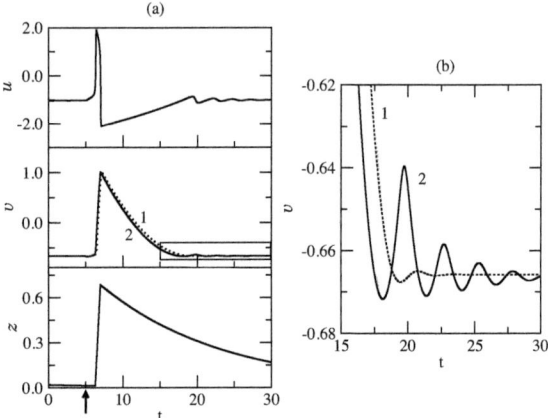

Figure 3.6: (a): Temporal evolution after perturbation in the single-unit model at $C = 0.008$. The spike is initiated by a short excitatory pulse at the time indicated by the black arrow. Dotted line (1) in the middle panel shows the y-time course for $C = 0$. (b): the enlargement of rectangular area in (a) shows the subthreshold oscillations after spiking. The dotted (1) and solid (2) lines illustrate the cases of $C = 0$ and $C = 0.008$, respectively.

over frequencies a spike train contains, changes significantly when noise intensity is increased, or the coupling to the exterior is enabled. That can be seen in the left column of Fig. 3.7 where we compare the densities of the unperturbed FHN model (case $C = 0$, given in grey) with the FNK model at $C = 0.03$ (black line). For all panels of the figure, the noise intensity is assumed to be small, so that the stochastic driving can be regarded as weak enough to not dominate the whole dynamics. For very low noise intensity, when only few spikes appear during the observation time, both models produce essentially the same shape of $S(\ )$ (not shown in figure). In the FHN model, further increasing of leads to the formation of a broad peak at zero frequency that moves rightward and reaches the position at ≈ 0.06 at $= 0.01$ (Fig. 3.7 (a)-(c)). It corresponds to a more regular firing due to coherence resonance [85, 110, 48].

The FNK model shows a similar power spectrum at very weak and at the final ($= 0.01$) noise strength, while the evolution of the spectra with increasing noise is different. Instead of a single broad peak at zero, two sharp peaks appear at zero frequency and at ≈ 0.075. The inspection of time courses shows, that the first

3.3 Local dynamics under the influence of noise

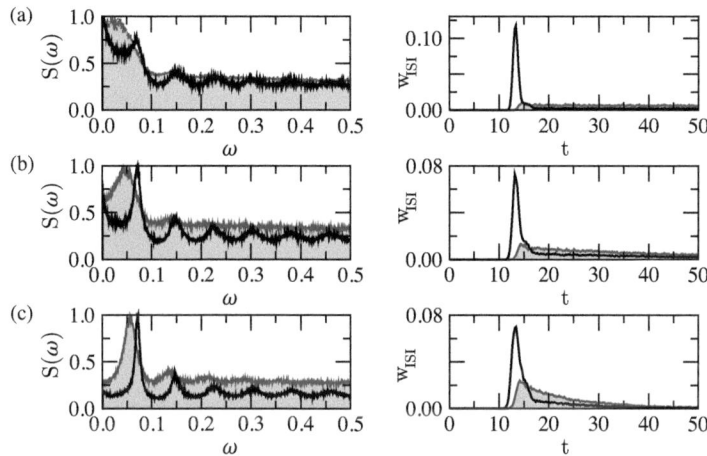

Figure 3.7: The spectral power density (left panels) and the probability distribution density of interspike intervals (right panels) for the single-unit model with noise. For comparison the maxima of the spectra are set to one. Curves in black were obtained with $C = 0.03$. Curves in gray with filled area were obtained with $C = 0$ illustrating the behavior of the unperturbed FHN model. Noise intensity takes values (a): $= 0.005$, (b): $= 0.007$ and (c): $= 0.01$ from top to bottom.

peak corresponds to randomly appearing spikes, while the second peak corresponds to the mean interspike distance within spiking events. With increasing , the peak at zero gradually disappears, while the second peak collects more power. The third row of panels in Fig. 3.7(c) shows considerably higher regularity of firing process in the FNK model compared to the FHN model.

Another quantity which is often inquired for spike trains is the distribution of interspike intervals w_{ISI}. It allows a statistical analysis of the times between two subsequent spikes. For the FHN and the FNK system it is shown in the right column of Fig. 3.7. While the activity near zero frequency is mapped on interval values with larger than 20 time units, a pronounced peak is observed at ISI values ≈ 1 which grows up to an optimal value with increasing intensity of noise strength

Both, the spectral power density and the interspike interval distribution are considered again in chapter 5 where their characteristics, depicted in Fig. 3.7, can be

43

3 Dynamical structures in a heterogeneous active medium

recovered in the abstract two–states model. The basis of that simplified approach provide waiting time distributions for rest and excited state, which are able to mimic the presented distributions for the complete FNK system.

To summarize the local dynamics, the behavior of a single unit is characterized by different dynamical regimes controlled by the parameter C that tunes the coupling to the additional equation for modeling the extracellular concentration z. It spans the mono–stable excitable behavior as known from the FHN system and multi–stable regimes containing stable periodic orbits or an additional stable fixed point at a high z-level. The addition of noise can lead to coherence resonance which is more pronounced than in the FHN case. The characteristics of this dynamics is reflected by both the spectral power density $S(\ )$ and the ISI distribution density w_{ISI}. The underlying mechanism of the discussed features can be regarded as subsequent self–induced depolarization enhanced by exterior potassium. These results are consistent with previously reported behavior of higher–dimensional quantitative models [64, 65, 37].

3.4 Two excitable units interacting with a common exterior

The self–depolarization described above plays an important role when two excitable cells share one z reservoir. Both units contributes to the rising of z when they are excited. In return they are also responsive to that increased concentration. We replace equation for the exterior Eq. 3.7 by:

$$\dot{z} = \left(\Psi(u_1) + \Psi(u_2)\right) - z, \qquad (3.10)$$

where u_1 and u_2 belongs to two separated units each described by the Eqs. (3.2–3.3) coupled to the common variable z. Depending on the coupling strength C firing of one unit can provide depolarization for both. We want to estimate how large C at least has to be in order to depolarize the second unit. Therefore, we consider the dynamics of z while one unit is excited and the other unit is at rest: $\Psi(u_1) = 1$ and $\Psi(u_2) = 0$. Then Eq. 3.10 can be integrated giving:

$$z(t) = -\left(1 - e^{-t}\right). \qquad (3.11)$$

As long as the unit is active z is delivered and the given expression describes an monotonic increase in z. The activation time is the time of allocating the stable cubic nullcline branch at $u > 0$ which can be approximated by linearization. It is determined by the parameters a_0 and   that we kept constant at $a_0 = 1.04$ and $= 0.04$ yielding an activation time of $t_{\text{act}} \approx 0.48$.

The lower fixed point of the second unit turns to be unstable, when the effective excitability parameter becomes smaller than one: $a_e = a_0 - Cz < 1$. Inserting

3.4 Two excitable units interacting with a common exterior

Eq. 3.11 into this inequality we find a minimal coupling strength:

$$C \overset{!}{\geq} -\left(\frac{a_0 - 1}{1 - e^{-t}}\right) \xrightarrow[t \to \infty]{} -(a_0 - 1). \quad (3.12)$$

That means, an elevated z–level lowers the excitability parameter value of the second unit for a certain time t. When the time window is long enough this parameter sinks below the critical Hopf–value and the second unit gets activated deterministically. Inserting the activation time of the first unit into Eq. 3.12 for which the z–value has reached its maximum we obtain a minimal coupling strength $C_{min} \approx 0.08$ (for = 1 and = 0.05). For any $0 < C < C_{min}$ the second unit exhibits subthreshold oscillations but cannot be activated without additional perturbations as illustrated in Fig. 3.8. Here, two coupling values are compared, one just below the minimal value (left column in the figure) and the other value exactly at the estimated number $C = C_{min}$ (right column). The activator evolution in time, marked with indices for the two cells and the common z–concentration is shown. An external perturbation

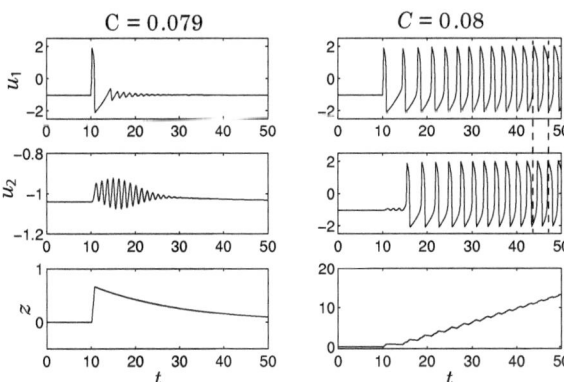

Figure 3.8: Activator evolution and common z–level for two coupled excitable units below and at the estimated minimal coupling value after a superthreshold excitation of the first unit. Left panel: The spike of the first unit leads to subthreshold oscillations in the second unit followed by an eventual decay into the rest state. Right panel: The first spike evokes an anti-phase firing sequence in both units.

initiates a spike in the first unit at $t = 10$ whereupon the z level rises, shown in the lowest row. That lowers the threshold of both units and causes subthreshold oscillations when the coupling is below the minimal value. Note the small range of u_2 in the figure. Or it triggers a self–sustained spike sequence when C is strong enough. In the latter case the units perform anti–phase oscillations after a transient (see dashed lines in the figure), an effect which is known from glycolytic oscillations in cells [159]. Perpetual firing let the z–level rise continuously. Following the same mechanism as described for Fig. 3.5 the system runs into the upper high z–level fixed point after some time where the oscillations die out, eventually.

The presence of noise can activate the second unit during the phase of subthreshold oscillations even below the minimal coupling values. Then the increase of z may be sufficient to evoke the self–sustained oscillation scenario. On the other hand, noise may also block repeated mutual spike induction for a coupling value greater than $C_{\min}$. Thus noise either acts constructive or suppresses activity, depending on the specific realization.

These considerations in that section will assist the understanding of the firing patterns in extended situation of many interacting units which follows in the next section.

3.5 Patterns in a spatially extended medium

Several ways are possible to create an extended scenario using the FNK model as the locally acting reaction dynamics from Eqs. (3.2–3.6). According to the initial biological situation we consider an inhomogeneous medium with separated active units, corresponding to disjointed cells embedded in the extracellular space being the diffusive medium for potassium ions described by z.

However, this is not the only option. Alternatively, we could assume that the $z(\vec{r}, t)$ variable describes a continuous diffusive medium with spatial coordinates $\vec{r} = (r_1, r_2)$. The excitable units would be placed in a second layer at locally separated sites coupled by the diffusing field $z(\vec{r}, t)$. This geometry is evident and leads to an usual reaction–diffusion system with three variables. Two variables are locally defined and coupled via the third. One might imagine a two layer system with excitable units located inside a gel with low connectivity. The interaction inside this first layer may have a much lower diffusion coefficient compared to the diffusive coupling of the third species $z(\vec{r}, t)$. For such mixed systems with densely packed excitable particles surrounded by reactive emulsions pattern formation has been observed in chemical experiments [26, 144].

Here, we focus on the first approach, whose geometry we call binary medium. It consists of immobile excitable elements embedded in the non–excitable field $z(\vec{r}, t)$, which diffuses in the remaining space of the two dimensional medium. In particular, we use a regular array of active units in each row and column illustrated in Fig. 3.9 as gray, that follow Eqs. (3.2–3.3). For each point of the intermediate diffusive

3.5 Patterns in a spatially extended medium

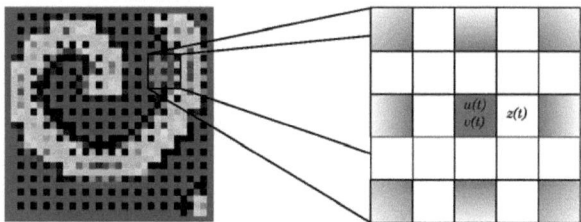

Figure 3.9: Schematic picture of the extended geometry illustrating the binary medium of neurons (N) and pure z-cells in between.

medium (white fields in Fig. 3.9) we write:

$$\dot{z}_{ij} = \sum_{k \in nb_1} \Psi(u_k) + \sum_{l,m \in nb_2} (z_{lm} - z_{ij}) - z_{ij}, \quad (3.13)$$

where the subscript ij denotes the current point in space. The second additional term describes the coupling within the z-field controlled by the coupling strength . The sum indices $nb_{1,2}$ denote the sets of neighboring units defining the coupling range. We have implemented several types of coupling such as nearest or next-nearest neighbor coupling for nb_1 and/or nb_2. To keep the dynamics close to the case of discrete diffusion we discuss the results only for the coupling to the next 8 surrounding boxes, where the diagonal elements are scaled by a factor of $1/\sqrt{2}$.

In contrast to conventional reaction–diffusion systems, the active units do not interact mutually but only via the common external concentration. The z-value that enters in the inhibitor equation of the active units is the average from the surrounding z-cells. The external medium is locally coupled with itself and is additionally affected by the neurons activity. All computations are performed in conditions in which identical individual units possess at least one stable fixed point at $u^0 \approx -a_0$. Patterns that will be presented in the following are formed from that non–active polarized state.

Except for the noise realizations the active units are chosen to be identical and parameters defining their dynamical behavior are fixed for all cases as: = 0.04, $a_0 = 1.04$, $C = 0.1$, $_l = _r = 1.0$, $u_s = 0.05$. Then they are in the excitable state supplied by a coupling to the external variable which potentially allows a rich dynamics due to multi–stability as known from Fig. 3.3. We examine the influence of parameters controlling the external medium according to the following table:

3 Dynamical structures in a heterogeneous active medium

parameters					z_{max}	boundary
set 1	50.0	6.0	2.0	0.00005	8.3	no–flux
set 2	60.0	6.0	130	0.02	10.0	absorbing
set 3	6.0	0.35	4.0	0.0001	17.1	periodic
set 4	10.0	0.5	0.2	0.00002	20.0	no–flux
set 5	150.0	6.3	2.0	0.003	23.8	periodic

With the increasing set number the mean z–level rises successively and with it the maximal z value as shown in the 6th column. According to the bifurcation diagram Fig. 3.3 we go to higher C values along the abscissa and encounter excitable, oscillatory and bistable behavior. We underline that the rest state of the uncoupled FHN is always a stable homogeneous state of our dynamics with $z = 0$. Every excitation of spatio–temporal structures presented here are evoked by noisy super–threshold stimuli.

3.5.1 Waves, spots and spirals (Fig. 3.10)

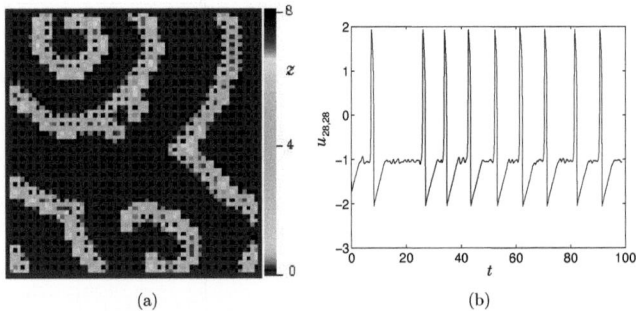

(a)   (b)

Figure 3.10: Parameter set 1: (a) Noise induced spirals and wave fronts. Gray level indicates the z–level and white and black squares represent active cells in the excited and rest state, respectively. (b) Activator time series of an arbitrarily chosen cell. (Colors online)

The local dynamics possess one single fixed point corresponding to the polarized state of neurons and is therefore excitable. By noise the units can be activated and release z to their neighborhood. The external concentration of the medium decays much faster than the units recover, while the diffusion is too slow to distribute the delivered z over a large distance. The activated units ignite next–nearest neighbor units via the medium and traveling extended waves are formed as it is depicted in Fig. 3.10 (a), where dark gray stands for a z level close to zero and light gray marks

3.5 Patterns in a spatially extended medium

the highest z concentration close to z_{max}. Black and white dots indicate the location of active units in their rest (black) and excited (white) state.

At the system borders or due to noise circular waves can break and the free ends curl forming a spiral wave. Although the dynamics is purely excitable waves appear very regularly at the chosen noise intensity, noticeable in the time series of an arbitrarily chosen unit, shown in Fig. 3.10 (b). This ordering effect of noise is also known from a homogeneous excitable medium [100]. The mean firing rate of the active cells is $r_{mean} \approx 0.1$ and the mean external concentration is $z_{mean} \approx 0.8$.

Slightly increased noise strength destroys the spiral wave structure by splitting it into short fragmented traveling segments that nucleate and annihilate in a random manner as described in section 2.3.3. The release of potassium in the exterior is still sufficiently low and no fixed point exists at high z-values. The external concentration has time to decay to zero during the refractory and inactive periodes of the units as indicated by the dominating dark gray in the figure.

For comparison, in the two–layer system and for a decay rate of ≈ 4 or smaller only short living wave segments supported by noise appear, indicating that this situation is close to the subexcitable regime.

3.5.2 Noise supported traveling clusters (Fig. 3.11)

The pronounced difference to the former set of parameters is the very large z–diffusion coefficient and a high noise level. Furthermore, it is the only set for which we set absorbing boundaries, where $z = 0$. Nuclei that would lead to coherent patterns like spirals diffuse very fast forming still connected clusters of delivered z. Such developed clusters can live relatively long wandering through the medium due to the forcing.

The local dynamics is excitable and possess only the low fixed point as a steady state. Due to the fast z–diffusion a large group of units gets activated whenever an unit is excited. It supports the formation of a localized high–level z region. Inside those clusters the threshold of the units is lowered which support the release of z. This process leads to self-feeding meandering cluster as depicted in the snapshots of Fig. 3.11 (c-f). Due to absorbing boundaries released z cannot accumulate. Diffusion quickly transports z to the boundaries where it flows off.

Note that noise is necessary during the whole time to keep the clusters alive. When the noise would be switched off the z–level would decay completely to zero and with it the clusters. Therefore we identify them as noise–supported.

For the given noise intensity the stochastically occurring spike events are quite regular. The z level follows the activation and forms also coherent oscillation–like elongations as shown in Fig. 3.11 (b). The mean rate is $r_{mean} \approx 0.2$ and the mean z concentration is $z_{mean} \approx 1.0$.

3 Dynamical structures in a heterogeneous active medium

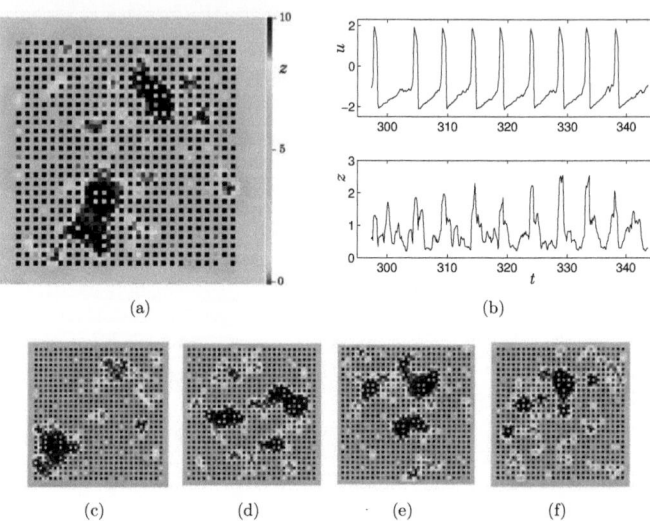

Figure 3.11: Parameter set 2, Gray level as in Fig. 3.10. (a): self–feeding clusters, (b): activator time series of an arbitrarily chosen cell and of a neighboring z–cell, (c)-(f): snapshots of nucleating, wandering and decaying clusters (Colors online)

3.5.3 Desynchronized oscillators embedded in a z-sea (Fig. 3.12)

Compared to the former case less potassium is released but it is also decay more slowly. The firing time of excitable units $t_{\text{spike}} \approx 0.5$ is shorter than the decay time $^{-1}$ of z. Therefore we observe oscillating units (Fig. 3.12) embedded in an exterior in which a high z–level survives longer than the duration of one oscillation period. Thus the exterior is permanently fed by potassium which is distributed quickly over the whole medium, shown in Fig. 3.12 (a) as homogeneous gray level.

Starting at the $z = 0$ level, the active units first perform the noise induced transition to the oscillatory behavior. Except for the the initiating perturbation noise is not needed to keep the oscillation alive. All units moves along the stable periodic orbit but with different phases. Along these units the medium is quickly filled with z which starts to propagate elevating the neighborhood and forming a front like spread over the space. It is a typical scenario of nucleation in systems with multiple attractors.

After the transition a quasi–steady picture remains with a sea of high potassium

3.5 Patterns in a spatially extended medium

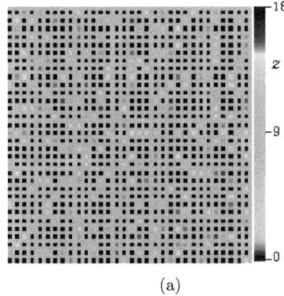 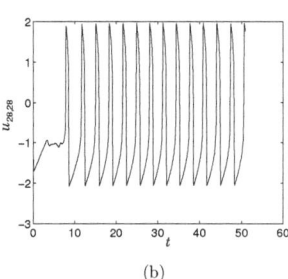

(a) (b)

Figure 3.12: Parameter set 3: (a) Elevated z–level due to permanently oscillating active cells. Gray level as in Fig. 3.10 (a). (b) Activator time series of an arbitrarily chosen cell. (Colors online)

populated by active units blinking regularly and feeding the exterior with potassium. We find for the oscillation frequency $r_\mathrm{mean} \approx 0.34$ and for the mean exterior $z_\mathrm{mean} \approx$ 5.7.

For a slower diffusion, after a transient, a chess–like blinking pattern is formed based on the anti–phase firing of neighboring units inside a sea of high z–level. In the deterministic case regular phase waves move across the medium. Noise adds irregularity and evokes wave break–up and desynchronization.

In Fig. 3.14 (a) the spatial correlation function is shown for an arbitrary unit over the distance to its neighboring excitable units along a row. The solid line represents the long range correlation to the active units in the neighborhood shortly after the wave–like propagation of the stimulus. A slow decay of the correlation can be seen expressing the indirect diffusive coupling. The first dip corresponds to the next-nearest active unit which is less correlated to the considered unit than the next but one. It reflects that on average neighboring elements fire preferable in anti–phase. However, a small amount of noise will drive the system to a complete desynchronized state after a couple of oscillations, shown as the dashed line in Fig. 3.14 (a).

The described situation is typical for the considered extended system and can be found over a large parameter range. Also in the two–layer system the same oscillating regime exists for the same parameter set.

3.5.4 Oscillations form a propagating ring-like pattern (Fig. 3.13)

Similar to the former parameter set, a single cell, fluctuating around the rest state, can reach the stable limit cycle by overcoming the unstable limit cycle due to noise. For the chosen noise level these events are rare. Once happened   is so small,

3 Dynamical structures in a heterogeneous active medium

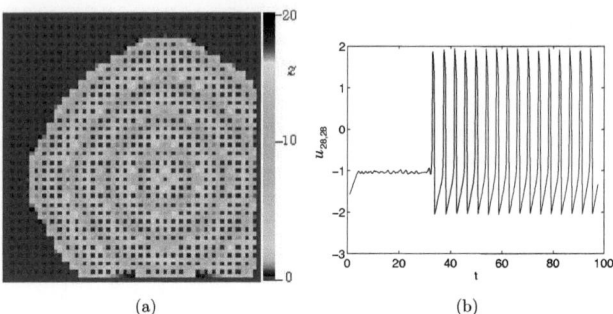

(a) (b)

Figure 3.13: Parameter set 4: (a) Noise induced concentric waves. Gry level as in Fig. 3.10 (a). (b) Activator time series of an arbitrarily chosen cell.(Colors online)

that the z–level can rise around the oscillating cell and lowers the threshold of the neighboring cells without decaying before. Therefore all active units can be elevated to the oscillatory behavior successively and a concentric wave appears, shown in Fig. 3.13 with a typical time series for u recorded from a single cell. For the chosen parameter values the oscillation period after the transition is $r_{\mathrm{mean}} \approx 0.25$, while the z–level averages $z_{\mathrm{mean}} \approx 6.0$. Compared to the last case diffusion of potassium is reduced drastically. This gives rise to the fact that the spatial structure can establish at the length scale of a few units.

In Fig. 3.14 (b) the spatial correlation is contrasted with set 3. The solid line shows a long range correlation shortly after initiating the wave pattern. The active units are well synchronized and it takes longer time until the structure is destroyed by noise. The latter desynchronized state corresponds to the dashed line in Fig. 3.14 (b).

Increasing   the stable and unstable limit cycle annihilate and the local dynamics is excitable with a single stable fixed point. A noise induced super–threshold perturbation leads to a singular firing event of the active cell and the z–level in its neighborhood increases. This elevated concentration ignites neighboring units once and a singular concentric ring–wave emerges.

Further increase of   shifts the dynamical behavior closer to the subexcitable regime in which wave segments with open ends exist. Such activated wave segments can be stabilized over a long time while they travel through the medium. Those patterns always decay in the two–layer situation. The released z can diffuse to each site of the array without restriction. Hence, the release rate   needs to be larger in order to provide the development of stable patterns. Choosing   = 15, for example, and starting with an initial nucleus which is large enough a structure is formed

3.5 Patterns in a spatially extended medium

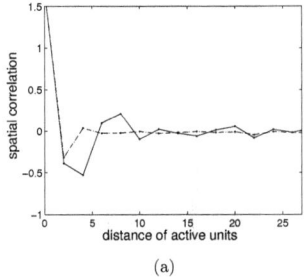 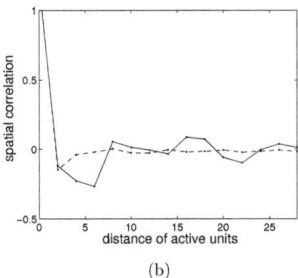

(a) (b)

Figure 3.14: Spatial correlation of set 3 (a) and set 4 (b) over the distance to neighboring units neglecting the inactive sites. Solid lines indicate transient behavior for early times, dashed lines show the system which becomes uncorrelated due to noise.

which we want to call oscillon. It is an extended but localized spot fed by released z supported by oscillating cells inside and surrounded by inactive cells. The fixed radius originates in the balance of release rate and diffusion versus the decay rate of z.

3.5.5 Bistability and inverted waves. (Fig. 3.15)

In this last considered parameter set we apply the largest release rate coming along with the highest z–level. After the nucleation of a bistable wave, as shown in Fig. 3.15 (a) the entire space becomes occupied by the high–level z state ($z \to z_{\max} \approx 24$).

The second stable fixed point exists due to bifurcation of a additional upper state corresponding to a constant depolarization of the units, as discussed in the section 3.2.3 for the local dynamics. The vicinity of the activated cells is permanently filled with z dissipated with the rate   and diffusively distributed with the coefficient   . Similar to the situation depicted in Fig. 3.6, during the transition from the low to high z state units oscillate within the propagation front. The extension of this oscillating front can be enlarged depending on parameter values.

Using periodic boundaries after the front has propageted through the medium the depolarized high z state is frozen. Biologically interpreted the medium is contaminated with potassium keeping the neurons permanently depolarized. That would lead to a serious damage of the cell tissue after few minutes and finally causes cell death as referred in the introduction of this chapter [36].

The presence of noise can break this frozen situation. Stochastic fluctuations may force single units into the transient oscillatory regime, so that the polarized lower

3 Dynamical structures in a heterogeneous active medium

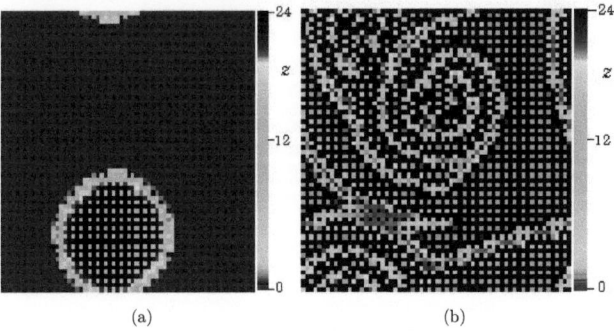

Figure 3.15: parameter set 5: (a) A bistable wave covers the medium with the high-level z state. Gray level as in Fig. 3.10 (a). (b) Noise induced inverted spirals and waves with polarized states appear. The z–scale is the same as in (a).(Colors online)

state is crossed and less z is released. That leads to a local decay of the maximal z state. This mechanism creates holes engraved into the high–level z sea. By diffusion those wholes can spread forming patterns like inverted spirals or propagating waves as shown in Fig. 3.15 (b). They can be seen as the symmetric relatives to the typical waves and spirals as found for set 1. Such coherent patterns do not exist in the vicinity of the considered parameter range for the two–layer system, after the high–level z state is reached. Only singular and disconnected holes occur stochastically.

Note that the lower state is still stable and the dynamics is bistable or even multistable. Thus, depending on the noise configuration, the system can reach the $z = 0$ state completely by the backward propagation of an inverted bistable front.

Other types of unconventional patterns have been reported previously. These are rotating spirals or target waves which run from outward to the center called antispirals or antiwaves, respectively. Such patterns have been found in the BZ reaction and other chemical reactions and can be described by specific reaction–diffusion systems [151, 130, 102]. However, in our case the rotational direction of the spirals and the propagation of waves is the same as for common waves as presented in Set 1.

3.6 Noise induced regularity at absorbing boundaries

The discovered diversity of patterns in the FNK model lead to several open questions. One of these concerns the interaction with boundaries. The latter parameter set 5 is

3.6 Noise induced regularity at absorbing boundaries

described for periodic boundaries. The situation changes, when fixed boundaries are used. In particular, we want to take a look at Dirichlet boundary conditions with fixed $z = 0$. In contrast to periodic boundaries, the medium cannot reach the frozen depolarized state after it is filled with high level potassium. The concentration flows across the boundary when a high z–front arrives at them. Under certain conditions and without noise the front turns around and moves back to the origin of the nucleus returning the whole system to the $z = 0$ level. We will discuss the interaction of a bistable fronts and the conditions of rebounding in more detail in the next chapter.

For the extended FNK model we study parameter values that let the front reflect on the boundary. We focus on the effects of noise which increases the probability of nucleation events. After the rebound of an initial front from the boundary the region near the boundary is prone to get excited by noise. Successive nucleations start spreading while the part of the font which is directed to the boundary gets reflected. Thus a wave train appears running from the boundary to the center. This is illustrated for a one–dimensional FNK system in Fig. 3.16 (a) where the z–level is shown in a space–time diagram. High level z waves (white) are created close to the boundary and move to the center more or less regularly, where they annihilate with the wave coming from the opposite boundary.

For an optimal chosen noise intensity a wave nucleates directly after the recovery time of the previous wave. Depending on parameter values the recovery time is of the order of (10) and thus the wave generation rate for this case is around 0.1. This value is almost reached as shown in Fig. 3.16 (b) where the nucleation rate is depicted in dependence on noise intensity and distance to the boundary. The rate diminishes for higher noise because waves can be destroyed which disturbs regularity. On the other hand less noise leads to gaps of irregular length after a wave is created. Thus, there is a maximum in the nucleation rate for an optimal noise value. Such effects where a finite noise intensity maximizes the regularity whithin a dynamical system where already mentioned for the patterns from set 1. This counterintuitive mechanism is often found in excitable system studied in numerous works as [100, 90, 63].

Additionally the dependency on the distance to one of the boundaries is shown. As it can be seen, the highest nucleation probability does not occur directly at the boundary but a little distant. This fact is due to the strict fixation of the boundary site which influence the nearest neighborhood of the boundary. Farther from the boundary the rate decreases rapidly because those sites are mostly occupied with traveling waves where further events are impossible.

Closing the section about spatially extended patterns we want to subsume the reported patterns in an overview given in Fig. 3.17. As already presented, the patterns are arranged in the order of rising z–level. For the second axis we choose the diffusion parameter which controls the pattern expansion. The left handed section represents essentially the mono–stable excitable medium close to the FHN dynamics. Thus we find the classical spiral and wave patterns. In the middle domain

3 Dynamical structures in a heterogeneous active medium

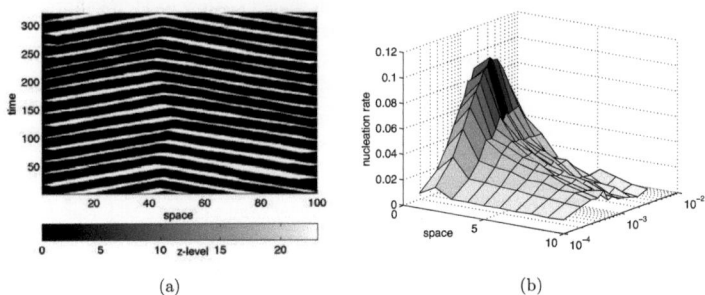

(a)   (b)

Figure 3.16: (a): Space–time plot of the one–dimensional FNK system interacting with absorbing boundaries. The z–variable is shown coded in gray levels.
(b): Nucleation rate in dependence on noise intensity and distance to the boundary. Parameter values: $= 0.04$, $a_0 = 1.04$, $C = 0.1$, $= 300$, $= 13$, $u_s = 0.05$, $_l = 6$, $_r = 1$, $= 4$ and $= 0.00175$

for intermediate z–level units possess a stable periodic orbit solution. Their self–sustained oscillatory behavior delivers z permanently in the environment. Typically, firing events are desynchronize, while embedded and supported by a vagrant z–sea. Extended wandering clusters appear through high diffusion, whereas low diffusion results in higher regularity and therefore coherent patterns. Even higher z stabilizes the upper steady state where inverted structures engraved into the potassium contaminated medium can occur. Note that boundary effects are not included in this scheme and therefore it is a rough classification.

3.7 Conclusions

In this section we have introduced a model that qualitatively describes the neuronal dynamics at variable extracellular concentration of potassium ions. Using the FitzHugh–Nagumo model as prototype for an excitable unit, we added a dynamical equation that qualitatively takes into account the potassium release from neuronal units and depolarization (threshold lowering) as a result of the increased extracellular potassium level. The analysis of the model for a single unit, for two coupled units, as well as for an extended array have shown that:

(i) The local deterministic model exhibits stable periodic orbits and an upper stable fixed point when the coupling C is increased in addition to the stable

3.7 Conclusions

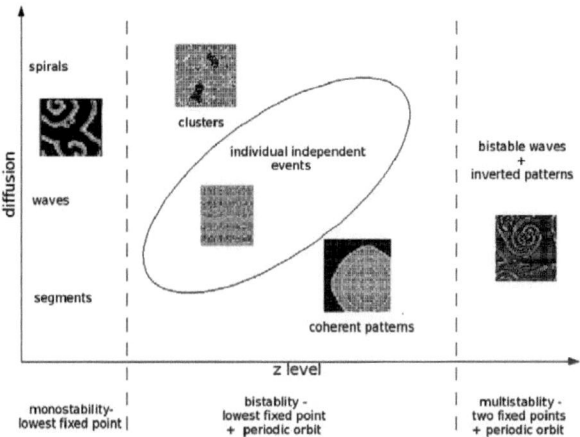

Figure 3.17: Overview of patterns presented in the current chapter ordered by the averaged z–level and the diffusion coefficient.

steady state from the FHN system.

(ii) The stochastic model of a potassium driven excitable neuron shows resonant behavior as subthreshold oscillations for higher values of potassium. With increasing noise intensity the power spectrum differs from the FHN type with strong time scale separation. It shows a pronounced and narrow peak at a frequency different from zero. The broad peak disappears with increasing noise intensity.

(iii) In the excitable regime two potassium–coupled neurons can ignite each other and trigger a spike sequence when C is large enough. The oscillation is synchronized in anti–phase and possessing a frequency doubled to the spiking of a singular neuron.

(iv) A two–dimensional array of potassium–driven neurons shows a variety of noise–induced spatial–temporal firing patterns depending on the release and decay rates of z and its diffusion coefficient. Besides patterns known from two–component RDS systems we find extraordinary patterns like long–living

3 Dynamical structures in a heterogeneous active medium

randomly–walking spots of depolarized states or the high–level potassium state with inverted spirals.

(v) For special parameter regions finite noise intensity leads to higher regularity in the nucleation rates close to absorbing barriers or in the local spike activity of active units.

In spite of the simplicity of the generalized model we use, some connections can be made between our results and relevant neuro–physiological studies, namely, the well known but still debated 'potassium accumulation hypothesis' [42, 32, 36] that considers the self–sustained rise of extracellular potassium as the cause of epileptic activity. Our computational results could be classified as following: the short–term activation of the z medium (including concentric and running waves) might describe the potassium dynamics within the physiological range, while the patterns with persistent high level z resemble the formation of epileptic seizure and thus can be regarded as representing pathological conditions.

Furthermore the presented patterns can be found in experiments of specific chemical reactions as already referred in the chapter. Typical patterns as waves and spirals found in set 1 and set 4 can be created for example in the in the BZ reaction [13] or in the CO oxidation on platinum [124] for which Gerhard Ertl receives the Nobelprize in chemistry in 2007.

4 Bistable wave fronts interacting with boundaries

4.1 Introduction

The previous chapter was closed with a discussion about nucleation effects near the system boundaries. In this chapter we want to extend these considerations by investigating the interaction of dissipative wave fronts with boundaries. In general, every imaginable experimental situation as well as every natural system is delimited by boundaries. Experiments in chemical laboratory conditions for instance are often implemented in petri–dishes or similar geometrical and material confinements. At those boundaries no chemical agents leak in the exterior which causes no–flux or Neumann boundaries defined by gradients of the concentration fields that vanish at the boundary position [82, 44, 46]. Corresponding reaction–diffusion models were numerical investigated in order to study the diversity induced by boundaries over an extended parameter space, which is not easily possible in experiments [109, 167].

So, it can be claimed, that for chemical reactions no–flux boundaries are the canonical boundary condition. However, there are biological situations for which other types of boundaries should be applied. Reminding the Potassium dynamics, as described in the previous chapter, we modeled ion release through a membrane into a extracellular region of low–level Potassium concentration. Neglecting detailed dynamics within this region the ions could be considered as getting absorbed immediately. That absorption could originate in a fast diffusion or in a highly receptive glia cell network. On the other hand one may consider the interaction of this low–level extracellular concentration with a passive bath of high–level concentration. In both cases the boundary values need to be fixed at a constant concentration value. These Dirichlet boundaries are often called absorbing boundaries, because in many cases boundary values are set to zero. For theoretical modeling of complex patterns in RDS absorbing boundaries are sometimes implemented in order to study the interaction mathematically [138, 7].

To not leave things incomplete we mention the two remaining main types of boundary conditions which are periodic and natural boundaries. The first deals with a finite geometry that is spatially folded at the begin and at the end. So the spatial geometry is a ring in one or a torus in two dimensions. The second type are boundaries that fix the field values in the infinity to make sure that there is no interaction of patterns with any system limits.

4 Bistable wave fronts interacting with boundaries

In the following chapter we are indeed interested in interaction effects of patterns with boundaries while we focus on the absorbing or Dirichlet type. We will approach that issue from different sides corresponding to different levels of assumptions and simplifications. However, one assumption every approach has in common. We concentrate exclusively on bistable fronts within this chapter. With this simplification we only need to consider a single transition from one phase to another, in contrast to excitable pulses that possess an additional back side. Considered bistable fronts here may be regarded as the forefront of an excitable wave in its short term behavior. Then it approximates a spatially extended excitable wave before the recovery period.

Bistable fronts are one the very first dissipative patterns that has been studied, because they already appear in one–component system that have at least a nonlinearity of quadratic order. The dynamics is described by the famous Fisher–Kolmogorov equation [33, 71]. Later on, other systems with traveling fronts are investigated for example for a reaction–diffusion model of an abstract bistable chemical reaction, for the spin propagation in the Ising-model or for crystal growth [127, 41, 78] Wave propagation is a phenomenon that occur far from the thermal equilibrium where effects of finite temperature causing stochastic fluctuations play an important role. Hence, the impact of thermic noise on front propagation attracted much interest [126, 104].

The model systems applied in the mentioned references are quite diverse. However, for the sake of manageability and because of its condensed mathematical structure some authors applied the FHN system in order to study universal phenomena of bistable fronts [45, 107]. As mentioned in the introductory section 1.1 the FHN model does not only exhibit excitatory or oscillatory behavior but also bistable dynamics. That is controlled by the relative position of the activator an inhibitor nullcline possessing three intersections in the bistable case. However, it is of eminent importance for the spatially extended case which slope the linear nullcline has. The effect on the bistable fronts is illustrated in Fig. 4.1 where the abstract phase space behavior and the corresponding space–time plot are shown. For the case shown in Fig. 4.1 (a) the inhibitor nullcline has a positive slope. The two fixed points, marked as black circles, are stable in the local dynamics. However, the addition of a diffusion term can destabilize them which lead to trajectories, sketched as arrows in the phase space figure, which run from such a diffusively destabilized fixed point to the other remaining stable fixed point. In space this is represented as a traveling front transforming the system from one state into another as it can be seen in the space–time plot. Fig. 4.1 (a) shows a situation in which the stability of both of the fixed points can fall prey to diffusion. Hence, fronts propagating in both directions can occur for one parameter set. For the attendant space–time plot the initial activator level is set to the upper fixed point except for a certain section where the level is set to the lower state. The edges of this inhomogeneity start to propagate in space and time as traveling fronts at first with different velocities and after a convergent transient with a common velocity and a finite distance. Note that front

4.1 Introduction

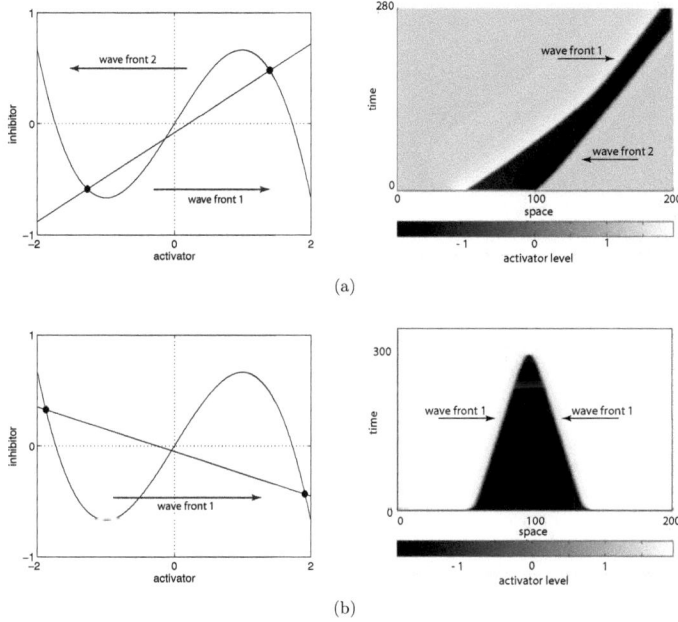

Figure 4.1: Two scenarios of bistable front propagation are illustrated. (a): A positive slope of the inhibitor nullcline can lead to diffusion–induced destabilization of the fixed points causing traveling fronts in both directions. (b): A negative slope of the inhibitor nullcline solely lead to a compensation of a initial inhomogeneity.

directions does not mean spatial directions here, but define the transition from the lower to the upper stable state or vice versa.

In Fig. 4.1 (b) the contrary case is shown, in which the stability of the two fixed point cannot be influenced by diffusion. An initial inhomogeneity, such as the anti-kink structure we have used for the first case, induces a front propagation that transforms the whole medium into that fixed point state, whose attractive force is stronger. This decision can be controlled by a slightly asymmetric position of the linear nullcline.

One of the characteristic quantities describing theses fronts is the front velocity,

61

4 Bistable wave fronts interacting with boundaries

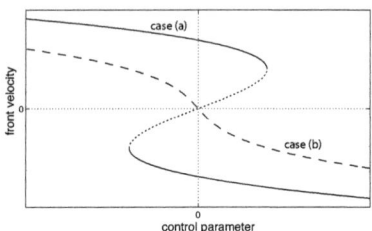

Figure 4.2: Scheme of two scenarios of the velocity dependence on a control parameter. The solid/dotted line represent the case in which two stable velocities can coexist whereas the dashed line represent a monostable velocity regime

which will play an important role in the following considerations. For this reason we want to touch an important feature of such bistable two–component systems even though it will be treated in detail in the coming sections. As suggested for the first scenario presented in Fig. 4.1 (a) two values for the front velocity with opposite signs coexists for a certain parameter range and thus the dependence of this velocity on a control parameter has a hysteresis shape. In Fig. 4.2 this case is contrasted to the second presented scenario from Fig. 4.1 (b) in which the velocity always depends monotonically on the control parameter. This general feature is referred also for front dynamics in different RDS [103]. In the following chapter we will focus on the more complex first case (a), shown as solid lines for stable velocities in the figure and a dotted line for the unstable velocity branch. Note least because of the analogy to the FNK system from the previous chapter whose specific positions of nullclines corresponds to this case.

At first we will introduce a local bistable form of the FHN model, with which we are already familiar. Before we study the effect of diffusion in a continuous spatially extended system we consider an one–dimensional array of discrete coupled units. That allows an investigation of the front velocity and boundary effects on the level of nullclines in phase space. Those discrete arrays are nothing completely abstract but can be investigated experimental and theoretical for chemical systems [80] or for coupled electronic units [91, 8]. The continuous bistable front will be studied for a diffusing activator and a locally acting inhibitor as well as for diffusion in both variables. The main quantity that characterizes the dynamical regimes we will study here, is the front velocity. Beside a relation of the velocity and the distance to boundaries we will also find approximate expressions for the bulk velocities and related critical parameter values indicating bifurcations. Finally we explore the influence of noise and its effect on the behavior close to the boundaries.

4.2 Local kinetics and linearization

For the specific form of the FHN system that will be used as the local reaction, we choose

$$\dot{u} = f(u) - v, \quad \dot{v} = g(u,v) \qquad (4.1)$$

$$\text{with} \quad f(u) = u - \frac{1}{3}u^3 \quad \text{and} \quad g(u,v) = au - v + b.$$

The parameter has its usual function to control the separation of activator and inhibitor time scale and is assumed to be always small ($\ll 1$). A value of b below the saddle–node bifurcation: $|b| < b_{sn} = \frac{2}{3}(1-a)^{3/2}$ and $a < 1$ ensures the existence of three fixed points. Two of them are stable, if the contiguous Hopf–bifurcation is also passed: $b < b_{\text{Hopf}} = \frac{2}{3} - a < b_{sn}$. The slope of the linear inhibitor nullcline is given by the parameter a whereas b determines the shift and thus the degree of asymmetry of the fixed points. The activator values are given by:

$$
\begin{aligned}
u_1^0 &= -r\cos\left[\tfrac{1}{3}\arccos\left(-\tfrac{3}{2}(1-a)^{-\tfrac{3}{2}}b\right) - \tfrac{\pi}{3}\right] \xrightarrow{b=0} -\sqrt{3-3a} & \text{(stable)} \\
u_2^0 &= -r\cos\left[\tfrac{1}{3}\arccos\left(-\tfrac{3}{2}(1-a)^{-\tfrac{3}{2}}b\right) + \tfrac{\pi}{3}\right] \xrightarrow{b=0} 0 & \text{(unstable)} \\
u_3^0 &= +r\cos\left[\tfrac{1}{3}\arccos\left(-\tfrac{3}{2}(1-a)^{-\tfrac{3}{2}}b\right)\right] \xrightarrow{b=0} \sqrt{3-3a} & \text{(stable)}
\end{aligned}
\qquad (4.2)
$$

with $r = 2\sqrt{1-a}$ and the order $u_1^0 < u_2^0 < u_3^0$. The roots of the cubic nullcline function $f(u)$ yield three branches that can be inverted and linearized separately:

$$u(v) = \begin{cases} -2\cos\left[\varphi(v) + \tfrac{\pi}{3}\right], & \text{for } u < -1 \\ -2\cos\left[\varphi(v) - \tfrac{\pi}{3}\right], & \text{for } -1 \le u \le 1 \\ 2\cos\left[\varphi(v)\right], & \text{for } u > 1 \end{cases} \qquad (4.3)$$

$$\approx \begin{cases} -\sqrt{3} - \tfrac{1}{2}v, & \text{for } u < -\tfrac{2}{\sqrt{3}} \\ v, & \text{for } -\tfrac{2}{\sqrt{3}} \le u \le \tfrac{2}{\sqrt{3}} \\ \sqrt{3} - \tfrac{1}{2}v, & \text{for } u > \tfrac{2}{\sqrt{3}} \end{cases} \qquad (4.4)$$

where $\varphi(v) = \tfrac{1}{3}\arccos\left(-\tfrac{3}{2}v\right)$. The nullclines in phase space and the linearized branches are shown in Fig. 4.3 for typical parameters $a = 0.2$ and $b = 0.05$.

4 Bistable wave fronts interacting with boundaries

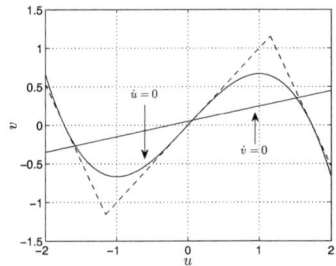

Figure 4.3: The nullcline of the Eqs. 4.1. Dashed lines represent the linearized pieces of the cubic nullcline according to the expressions in Eq. 4.4.

4.3 Array of coupled units

4.3.1 Thin front

We combine the bistable units to an one–dimensional array with a locally acting coupling term in the activator. In this discrete representation the model from Eqs. 4.1 can be rewritten as

$$\dot{u}_i = f(u_i) - v_i + \epsilon (u_{i-1} + u_{i+1} - 2u_i) \tag{4.5}$$
$$\dot{v}_i = g(u_i, v_i), \tag{4.6}$$

where i denotes the position of a unit and $\epsilon = D/h^2$ is the coupling constant in the discrete system which is the diffusion constant D in the continuous system in the limit $h \to 0$. We use Dirichlet boundary conditions for the right end of the array at the lower local fixed point at $u_{bR} = u_1^0$ and the left end at the upper fixed point at $u_{bL} = u_3^0$. Consequently, the initial condition for the array has to be inhomogeneous to fulfill these boundary conditions. In the following we set the left handed part of the units into the upper fixed point and the right handed part in the lower fixed point, respectively, as illustrated in Fig. 4.4. The direction of the front propagation, marked with the arrow in the figure, is due to the choice of values for a and b.

Assuming the front as a solution of a one–component bistable system for a moment, we can approximate the width of the front as $l_w = 4\sqrt{6h}/(u_3^0 - u_1^0)$ [126]. For the array of discrete units we can thus estimate an upper limit for the coupling constant ϵ_{max} below which the front consists of one single element, given as $\epsilon_{max} = (u_3^0 - u_1^0)^2/24$. The dynamics of this thin front can be investigated by considering the single front unit u_F having neighbors that are assumed to be fix at $u_{F-1} = u_L$ and $u_{F+1} = u_R$. That leads to an effective shift of the activator nullcline

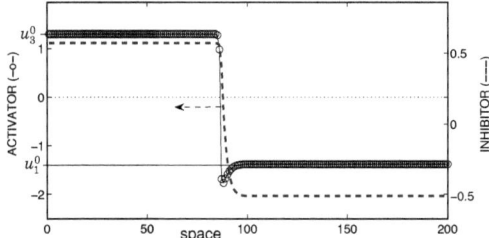

Figure 4.4: Set of units forming a bistable front according to Eqs. 4.5–4.6. The solid line with circles represents activator elements u_i and dashed line indicates the corresponding inhibitor values. Parameter values are $a = 0.4$, $b = 0.05$ and $\varepsilon = 0.1$ such that the front consists of a single unit. Dirichlet boundaries are chosen such that $u_{bL} = u_3^0$ and $u_{bR} = u_1^0$. This is the generic setup for fronts during the following chapter. Initial conditions can vary, however, to trigger both moving directions.

for the front unit due to an additional next–neighbor coupling term:

$$v_a = u - \frac{1}{3}u^3 + \gamma(u_L + u_R - 2u). \qquad (4.7)$$

The coupling is isotropic so that for $u_L - u = u - u_R$ the coupling term vanishes and the nullclines coincide with the uncoupled FHN model, see Fig. 4.5, solid curve 1. Relative to that unperturbed reference nullcline and to a fixed inhibitor line with $a = 0.4$ and $b = 0$ two special cases are illustrated in the figure. The dashed nullcline corresponds to the shift due to an unidirectional coupling to a neighboring unit which is elongated to the maximal activator level u_{max}. For a coupling value of $\gamma = 0.1 < \gamma_{max} = 0.3$ the lower intersection of the nullclines is shifted to a location where the corresponding fixed point becomes unstable. The same happens to the upper fixed point when for example the right handed and left handed neighbors are at or close to rest states and the coupling value chosen as $\gamma = 0.187$ (dashed–dotted line in Fig. 4.5). A unit that is affected by such discrete coupling constellations starts to move and can trigger the propagation of the whole front. This will be discussed in the next section.

In numerical simulations of Eqs. 4.5–4.6, using a spacing of $h = 1$ and an array that consists of 200 units, we find different modes of front motion and behavior at the boundaries. In Fig 4.6 four typical space–time plots of bistable fronts are presented. The labels 'region II - V' define separate parameter regions and will be discussed in more detail later. The label 'region I' will be referred later for a regime

4 Bistable wave fronts interacting with boundaries

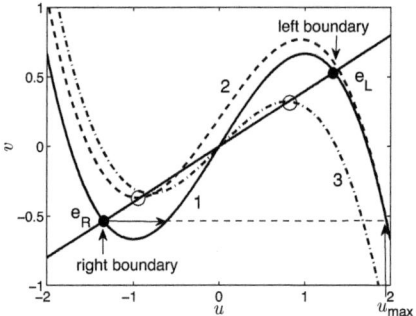

Figure 4.5: Nullclines according to Eqs. (4.6) and (4.7). Filled and open circles denote the stable fixed point locations states and states near Hopf–bifurcation, respectively. Boundary values are denoted by e_R (right) and e_L (left). The cubic nullcline of the front unit u_F from Eq. 4.7 is shifted due to elongated positions of neighboring units: The solid curve 1 refer to the free case (= 0). Dashed curve 2 corresponds an effective cubic nullcline of the front unit with neighbors at $u_{F-1} = u_{\max}$ and $u_{F+1} = u$ with = 0.1 (mono-directional coupling, lower fixed point becomes unstable). Dashed–dotted curve 3 corresponds to neighbors at $u_{F-1} = 1.3$ and $u_{F+1} = u_{bR} = -1.341$ with = 0.187 (interaction with boundary, upper fixed point becomes unstable). Parameter values $a = 0.4$ and $b = 0$ remain constant.

of no front propagation. As the initial configuration for the simulation half of the units are set in the upper activator fixed point value but only the first third of them have also the inhibitor value in the fixed point to force front propagation to the right. The remaining units are set in the lower fixed point. The parameter values for the slope of the inhibitor nullcline and for the time–scale separation are fixed at $a = 0.4$ and = 0.05, the regions though are defined by different values of and b. In three cases of the presented four examples in Fig 4.6 this initial setup actually leads to front propagation onto the right handed boundary whereas in the first case, labelled with 'region II' the front holds its position until the inhibitor front have reached the upper fixed point as well whereupon the fronts moves to the left.

The phenomenology of this and the remaining three cases are characterized by different boundary interactions, that will be quantified in the following sections.

4.3 Array of coupled units

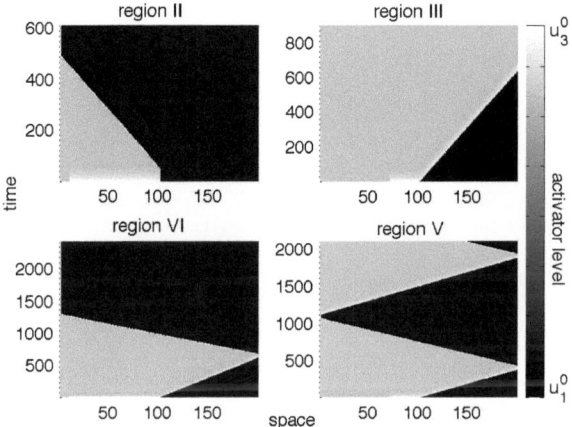

Figure 4.6: Different front behavior illustrated in space–time plots for varying parameters obtained by numerical simulations. The gray scale represents the activator value that ranges between the upper (white) and lower (black) fixed point. Parameter values for region II: $b = 0.15$, $ = 0.1$, region III: $b = 0.03$, $ = 0.15$, region VI: $b = 0.1$, $ = 0.2$, region V: $b = 0.03$, $ = 0.25$, remaining parameters for each region: $a = 0.4$, $ = 0.05$.

4.3.2 Propagation failure

In such reaction–diffusion systems with a discretized diffusion term there is always a coupling value, below which no signal can propagate. This phenomenon is called propagation failure [68, 80]. To estimate this value, let us assume that the coupling term $(u_{i+1} + u_{i-1} - 2u_i)$ should provide strong enough forcing such that an elongation of a neighboring site $u_{i\pm 1}$ induces a super–threshold activation in the considered unit u_i.

We expect the front to start moving when the coupling is strong enough to shift the fixed points (marked as e_L and e_R in Fig. 4.5) beyond the extrema of the cubic activator nullcline, where they are unstable. The maximal activator value, a unit

4 Bistable wave fronts interacting with boundaries

can reach, is given by

$$u_{max} = 2\cos\left(\frac{1}{3}\arccos\left[-\frac{3}{2}(au_1^0 + b)\right]\right) \tag{4.8}$$

$$\approx \frac{2\sqrt{3}}{a+2}(a - b + 1), \tag{4.9}$$

illustrated by a dashed line in Fig. 4.5. The expression in Eq. 4.9 is due to the linearized nullclines, mentioned in section 4.2. The next–neighbor coupling term $(u_{max} - u)$ acts on the considered unit with the strength   that can be large enough, to force the rest state into an unstable region. At the critical value for   , marked in Fig. 4.5 as empty circles, a subcritical Hopf-bifurcation takes place and an unstable periodic orbit appears. From the condition for the Hopf-bifurcation we obtain:

$$v_a' = (1 - \text{Hopf}) - u^2 \stackrel{!}{=} 0 \Rightarrow u_{\text{Hopf}} = \pm\sqrt{1 - \text{Hopf}}$$

$$v_a(u_{\text{Hopf}}) = v_i(u_{\text{Hopf}}) \Rightarrow \pm\left(\frac{2}{3}(1 - \text{Hopf}) - a\right)\sqrt{1 - \text{Hopf}} + \text{Hopf}u_{max} - b = 0. \tag{4.10}$$

At these critical parameter values, the lower fixed point e_R becomes unstable and the wave starts moving to the right according to the chosen initial conditions. The analogous consideration can be made for the destabilization of the upper fixed point e_L at which the term $(u_{min} - u)$ acts locally on the activator nullcline and makes the front run to the left boundary. Using the linearized terms for u_{max} and u_{min} we find two expressions $b_{1,2}(\text{ Hopf})$, shown in Fig. 4.8 as dashed lines. The monotonically decreasing line marks the transition from a not moving wave profile to a front propagating to the left, corresponding to the destabilization of the upper fixed point. For positive b this propagation failure limit is at lower   than for the front moving to the right (increasing dashed line). This can be understood by the asymmetric positions of the lower and upper fixed point for $b \neq 0$. The coupling–induced shift of the cubic nullcline thus may be only sufficient to destabilize one of the fixed points which makes the propagation failure dependent on the direction of motion.

4.3.3 Interaction with the boundaries

When the traveling front approaches one of the boundaries it either reverses and moves back in the opposite direction or it stops at the boundary and forms a stationary profile, as the four cases from Fig. 4.6 suggest. In the following we want to understand the mechanism of that different behavior and we will estimate parameter values that define the regions I – V.

We consider the front running to the left boundary where the boundary value (or the value of the zeroth unit) is set to $u_0 = u_{bL} = u_3^0$ and so is every unit before

4.3 Array of coupled units

the unit approaches. For a coupling $< $ max the units become elongated from the u_3^0 state one after another. Trajectories in phase space for the very first and second unit next to the boundary is shown in Fig. 4.7 in relation to the nullclines of an uncoupled FHN unit. The second unit almost perform the whole excursion in phase space as known from a uncoupled unit whereas the dynamics of the first unit is biased due to its fixed neighbor. Depending on the coupling strength, both units reach a stable state and remain in that new position as depicted in Fig. 4.7 (a) for = 0.24181. This local behavior corresponds to a front that stops at the boundary and forms a stationary profile. A coupling strength which is increased by an amount of 10^{-5} leads to trajectories that follow a homoclinic orbit returning to the upper fixed point (Fig. 4.7 b), corresponding to rebounding wave front.

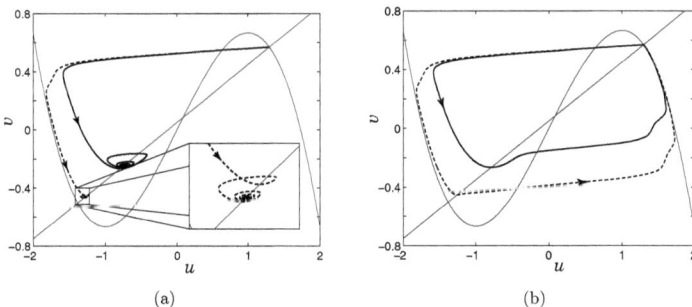

Figure 4.7: Excursions in phase space of the first unit (u_1, v_1) (solid line) and the second unit (u_2, v_2) (dashed line) of the FHN chain close to the left boundary.
(a): The front get stuck and a stationary front profile is formed for = 0.24181. The zoomed region shows the trajectory of (u_2, v_2) spiraling into the new fixed point. (b): The wave front rebounds at the boundary for = 0.24182. (remaining parameters set as: $a = 0.4$, $b = 0.05$ and = 0.05)

In order to investigate the critical value of we assume, that the front affects the units strictly separated. Then the next–nearest boundary unit u_1 has its neighbors in the moment of a passing front at the activator values: $u_0 = u_{bL}$ and $u_2 = u_1^0$. Note that this assumption underestimates the position of u_2 even for the thin front regime. As depicted in Fig. (4.7) that unit does not reach the lower fixed point, but it is close by, so that we approximate the coupling term for u_1 as $(u_3^0 + u_1^0 - 2u_1)$. We already mentioned the emergence of a Hopf–bifurcation to explain the appearance

69

4 Bistable wave fronts interacting with boundaries

of propagation failure. The same argument can be applied here for the transition of a stationary front profile at the boundary to a rebounding wave front. Similar to the condition in Eq. 4.10 we obtain

$$\pm \left(\frac{2}{3}(1 - 2\varepsilon_{\text{Hopf}}) - a\right)\sqrt{1 - 2\varepsilon_{\text{Hopf}}} + \varepsilon_{\text{Hopf}}(u_3^0 + u_1^0) - b = 0, \qquad (4.11)$$

where the positive sign belongs to the upper fixed point and the negative sign to the lower fixed point, respectively. The condition in Eq. 4.11 leads to two branches shown in Fig. 4.8 as solid lines. The first monotonically decreasing curve corresponds to the positive sign and thus to the transition from sticking front at right boundary to a reflecting front (region III – region IV). Within the region IV, beyond this line the upper fixed point become unstable and the front rebounds at the right boundary, however stops at the left boundary with $u_{bL} = u_3^0$. This asymmetry would be exactly reversed when the sign of the parameter would be changed. For ε a beyond the second solid line in Fig. 4.8 both fixed points are destabilized due to the local coupling and the wave repels from the right and the left boundary performing a continued oscillation between them.

Altogether we distinguish five regions in the bifurcation diagram Fig. 4.8. The parameter domain labelled as I is the region of complete propagation failure, in which the front does not move. In region II, the front can only move in one direction onto the left boundary where it gets stuck. In the small region III, the front is able to move to the left and to the right without rebounding at one of the boundaries. Rebounding on the right boundary occurs in region IV and in V the wave rebounds at both sides.

We compare the transition curves evaluated from Eq. 4.10 and Eq. 4.11 with numerical simulations of the Eqs. 4.5–4.6 represented by open circles in Fig. 4.8. The remaining parameters are fixed at $a = 0.4$ and $\varepsilon = 0.05$. The additional line consisting of stars marks the location of ε_{max}; the limit of the thin front approximation. The mentioned systematic underestimation can be noticed by the shift to smaller values of ε of the numerical curves. The main reason for that is the assumption of fixed neighbors in the coupling term. That is only true for $\varepsilon \to 0$ where the difference between the analytical and numerical results vanishes. For a coupling strength greater than ε_{max} the numerical curve clearly deviates from the analytical prediction. Here, more than one unit play a role for the front dynamics and the simple assumption of a coupling term with fixed neighbors fails.

4.3.4 Self–sustained oscillation near the boundary

The effects described in the previous section are governed by subcritical Hopf–bifurcations. Essentially, the front rebound is explained by the destabilization of stable fixed points due to the influence neighbors, which are assumed to be constant. For stronger coupling $\varepsilon > \varepsilon_{\text{max}}$ multiple units form the front shape and thus the neigh-

4.3 Array of coupled units

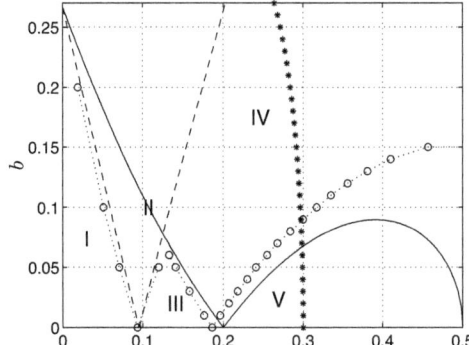

Figure 4.8: Bifurcation lines on (ϵ, b) parameter plane. Roman numbers indicate regions in which the front behaves such as displayed in Fig. 4.6. Dashed lines mark the limit of propagation failure to the left (decreasing line) and to the right (increasing line) evaluated by Eq. 4.10. Solid curves represent Hopf–bifurcation lines from Eq. 4.10 that separate regions of wave rebound only at the left (region IV) or on the left and right boundary (region V). Stars mark the limiting coupling ϵ_{max} below which one single unit forms the front. Circles show these transitions due to numerical simulations. Remaining parameter values are $a = 0.4$ and $\epsilon = 0.05$.

bors in the coupling term of these front units are definitively time dependent. Even in the continuous case the wave still either rebounds at the boundaries in the region of small b or gets stuck for all times.

If the wave profile at the boundary consists of multiple units, new fixed points can appear along the inhibitor nullcline, which are not necessarily stable (see Fig. 4.9). Close to the Hopf–bifurcation after which the front would rebound, small oscillations of the units belonging to the front near the boundary give notice of the close criticality. We will find such oscillating fronts at the boundary also for the continuous bistable front even more pronounced.

4 Bistable wave fronts interacting with boundaries

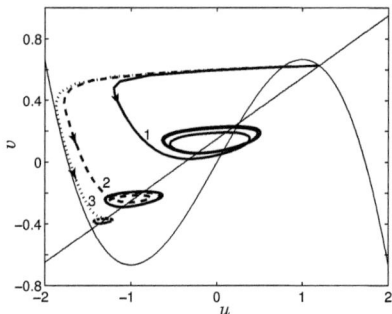

Figure 4.9: Excursions in phase space of the first unit u_1 (solid line), the second unit u_2 (dashed line) and the third unit u_3 (dotted line) of the array close to the left boundary. For $= 0.6$ and $b = 0.15$ the wave front gets stuck and a front profile is formed where units close to the boundary perform self–sustained oscillations. (Remaining parameters: $a = 0.4$ and $= 0.05$)

4.4 Continuous bistable front

In this section we turn to a continuous bistable medium described by partial differential equations. Two cases will be distinguished. First, the pure activator diffusion with a locally acting inhibitor corresponding to a situation similar to the biological background of the FNK model. There the inhibitor was interpreted as the immobile membrane or glial dynamics whereas the activator described the diffusing ion concentration. For this scenario no boundary effects will be discussed. However, we present an approach to treat such type of spatially extended dynamics. In the limit of small front velocities as well as for the limit of small time scale separation of activator and inhibitor evolution we will find approximate front shapes and critical parameter values where elemental bifurcations take place.

The second considered case is spatial diffusion of both, the activator and inhibitor. This may correspond to chemical reaction scenarios where the chemical concentration acting as an activator and the concentration acting as an inhibitor can spread in space. For this case we also find the elemental bifurcation even more precisely and furthermore we will be able to treat boundary effects. We will focus on Dirichlet boundaries that limits the one–dimensional space on both sides. Front propagation within this finite region shows different interaction behavior that we will quantify with an distance–velocity relation. This relation also holds for no–flux boundaries when one single sign is changed.

4.4.1 Diffusing activator – immobile inhibitor

Co–moving frame

In the first instance we present the method applied to investigate the following partial differential equations. For an activator concentration diffusing in a homogeneous active medium we obtain the one-dimensional reaction–diffusion equation:

$$\dot{u} = f(u) - v + D\frac{\partial^2}{\partial x^2}u, \quad \dot{v} = g(u,v), \qquad (4.12)$$

with the cubic function as exerted in Eq. 4.1. We use the same parameter set for the local dynamics a, b and so that the system is bistable. The discrete units from the previous section can be considered as densely packed in space with an infinitesimal small spacing ($h \to 0$). Beside the trivial homogeneous solution for the Eqs. 4.12 describing a state in which the whole system occupies one of the local stable fixed points and where the diffusion term vanishes, there is also a non–trivial inhomogeneous solution including a smooth front forming the transition from one stable fixed point to the other.

The aim of the following discussion is to quantify approximately that front and to clarify the different types of motion that exist in dependency on the diffusion coefficient and the local parameters. Therefore we change into co–moving front coordinates $\quad = x - ct$, where c denotes the velocity of the front. Thus both differential equations become ordinary and read

$$Du'' + cu' + f(u) - v = 0, \quad cv'(\ ) + g(u,v) = 0, \qquad (4.13)$$

where the prime stand for the derivative with respect to the new coordinate now. The quantity that characterizes the front motion is the velocity c. It may take negative as well as positive values, depending on the orientation in space and on the balance of the local fixed points. We will also find a stationary case with zero velocity and a front forming a stable profile. As mentioned in the beginning we assume a natural boundary scenario. That means, that not until $x \to \infty$ the front reaches the lower fixed point e_R asymptotically and for $x \to -\infty$ the upper fixed point e_L, accordingly. In fact, the front moves in space without interacting with any boundary in a finite time. Essential bulk properties such as the existence and stability of back– and forward motion can be studied that teach us the basic dynamical behavior helping to understand near–boundary effects.

4 Bistable wave fronts interacting with boundaries

Series expansion in c

The condition for the analysis of Eqs. 4.13 for small velocities is the existence of a solution where $c = 0$, labeled as u_0. Since there is no propagation failure in continuous systems we find such a stationary solution only for $b = 0$, where the lower and upper stable fixed point are symmetrical attractive (see Eqs. 4.2). Then the basins of attraction have the same size and no direction of motion is preferred. Around that stationary solution we expand the system in powers of c:

$$u(\ ) = u_0(\ ) + c\, u_1(\ ) + c^2\, u_2(\ ) + \quad (c^3),$$
$$v(\ ) = v_0(\ ) + c\, v_1(\ ) + c^2\, v_2(\ ) + \quad (c^3), \quad (4.14)$$
$$= {}_0 + c\, {}_1 + c^2\, {}_2 + \quad (c^3).$$

We substituting these expressions into Eqs. 4.13 and evaluate terms with the same power of c. This yields for the **0. order**:

$$D u_0'' + u_0 - \frac{1}{3} u_0^3 - v_0 = 0, \quad v_0 = a u_0. \quad (4.15)$$

The inhibitor follows immediately the activator dynamics and one ordinary differential equation with a cubic nonlinearity remains. Besides the homogeneous solutions corresponding to the three fixed points there is a non-homogenous 0. order front solution given explicitly as

$$u_0(\ ) = \frac{2 u_3^0}{1 + \exp(\pm\ )} - u_3^0 = u_3^0 \tanh\left(\mp \frac{}{2}\right) \text{ with } = \sqrt{\frac{2}{3D}} u_3^0, \quad (4.16)$$

as known from one-component bistable systems [127, 126].
Both given signs solve Eqs. 4.15. However, due to the boundaries the wave has to reach asymptotically, the sign can be fixed to be minus. This expression represents a stationary bistable front as an odd function of the coordinate which can be replaced by x because $c = 0$. The activator front thickness can be estimated as $l_w = \sqrt{8D/(1-a)}$.

The **1. order** in c produces the following differential equation:

$$D u_1'' + \left(1 - a - u_0^2\right) u_1 = \left(\frac{a}{{}_0} - 1\right) u_0'. \quad (4.17)$$

We identify $u_1 = u_0'$ for which the right hand side vanishes. Hence, the corresponding eigenvalue to u_0' that belongs to the differential operator acting on the left hand side is zero. Then the inhomogeneity on the right hand side is orthogonal to u_0'. That

4.4 Continuous bistable front

results in a condition for ϵ_0

$$\left(\frac{a}{\epsilon_0}-1\right)\int_{-\infty}^{\infty}(u_0')^2\,dx=0 \quad\Rightarrow\quad \epsilon_0=a. \tag{4.18}$$

At $\epsilon=\epsilon_0$ a supercritical pitchfork bifurcation takes place, where the stationary front solution becomes unstable. By substituting the result $\epsilon_0=a$ into Eq. 4.17, it becomes homogeneous and can be solved by choosing $u_1=0$. In the series expansion for $b=0$ that means, that the linear term always vanishes. To find stable propagating front solutions below the bifurcation value, it is necessary to evaluate higher orders of the expansion. From the 2. and 3. order in c we obtain the conditions:

$$\int_{-\infty}^{\infty}\left(u_0'u_0''' - \epsilon_1(u_0')^2\right)dx = 0 \quad\Rightarrow\quad \epsilon_1 = 0,$$

$$\int_{-\infty}^{\infty}\left(u_0'u_0'''' - a\,\epsilon_2(u_0')^2\right)dx = 0 \quad\Rightarrow\quad \epsilon_2 = -\frac{2}{5}\frac{(1-a)}{aD} < 0. \tag{4.19}$$

With the vanishing first order in the activator $u_1=0$ the same order is omitted in the v equation in the expansion 4.14. In the $c-\epsilon$ relation we obtain a characteristic square-root like behavior close the critical point and the front velocity below the bifurcation value is given by

$$c = \pm\sqrt{\frac{\epsilon-\epsilon_0}{\epsilon_2}} = \pm\sqrt{\frac{5}{2}\frac{a-\epsilon}{1-a}aD}. \tag{4.20}$$

An example for $a=0.1$ and $D=1$ is shown in Fig. 4.11 (a) where the solid line represents the stable velocity branches given by Eq. 4.20. Numerical results, shown as circles, are in good agreement close to the critical value at $\epsilon_0=0.1$. It is important to note, that for any $\epsilon > \epsilon_0$ no moving front solution exists whatever initial condition or remaining parameter set is used. Within this parameter region, after a transient time, which indeed depends on the latter, the activator and inhibitor fronts are completely balanced and stop moving. This is illustrated in Fig. 4.12 (a) as a space-time plot, where the inhibitor front is initiated in relation to the activator front such that in the beginning a motion to the right is induced. That regime may resemble the effect of propagation failure. However, the origin of the latter lays in the discretization of the array and thus does not exist in our spatially continuous system due to the characteristic $c \propto \sqrt{D}$ dependence verified in Eq. 4.20.

The corrections to the activator and inhibitor field until the 2. order in c read

$$u(\xi) = u_0(\xi) + \frac{c^2}{2aD}u_0'(\xi) + \mathcal{O}(c^3),$$

$$v(\xi) = au_0(\xi) + c\,u_0'(\xi) + c^2\left(\frac{1}{a}u_0''(\xi) + \frac{1}{2D}u_0'(\xi)\right) + \mathcal{O}(c^3). \tag{4.21}$$

4 Bistable wave fronts interacting with boundaries

By inserting the first expression into the second at the front position ($\xi = 0$) the parametric phase trajectory is given by

$$v(u) = au + \frac{c}{2u_3^0}\left(1 + \frac{c}{au_3^0}u\right)\left(u^2 - (u_3^0)^2\right). \tag{4.22}$$

The front shape in space from Eqs. 4.21 of both variables is compared with numerical simulations in Fig. 4.10 (a) for parameter values ($a = 0.3$, $\epsilon = 0.2$ and $D = 1$) that lead to a front velocity of $c \approx 0.33$. Whenever $c \neq 0$ the propagating inhibitor front lags behind the activator front. For first order corrections the bias of the fronts can be approximately written as

$\xi_{lag} \approx 2/\epsilon \; \text{arcoth}(c/\epsilon(a - \sqrt{a^2 + \epsilon^2 c^2})$. It grows with increasing c and goes to zero for vanishing velocity. The lag can be estimated for higher orders more precisely, but not in a closed form. Due to the bias of front positions, one can consider the inhibitor front to shift the activator front forward. In phase space this is represented as curved heteroclinic orbits connecting the two stable fixed points, illustrated in Fig. 4.10 (b). In contrast, for $c = 0$ the trajectory in phase space follows the inhibitor nullcline which can be seen directly in Eq. 4.22 when neglecting the second term on the right hand side.

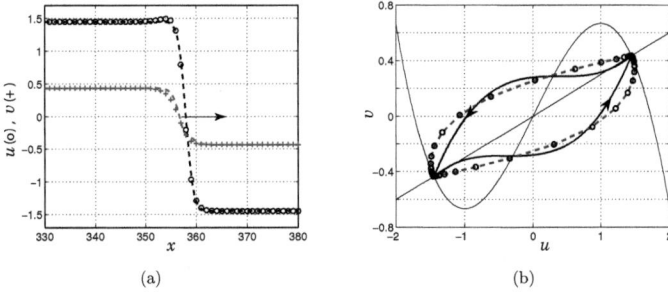

(a) (b)

Figure 4.10: (a): Shape of propagating activator (black and circles) and inhibitor (gray and pluses) fronts. Dashed lines are due to Eqs. 4.21. They are compared to numerical simulations of Eqs. 4.12 (symbols). (b): Corresponding trajectory in phase space at $\epsilon = 0$ with nullclines (thin lines). Numerical simulations (symbols) compared to the analytical expression from Eq. 4.22 (thick solid line). Parameter values: $a = 0.3$, $b = 0$, $\epsilon = 0.2$ and $D = 1$.

4.4 Continuous bistable front

Series expansion in

Up to this point the approach is valid close to the stationary solution with $c = 0$. The condition was a symmetric situation with $b = 0$. A supercritical pitchfork bifurcation point marks the transition from one stable stationary front solution to an unstable stationary front attended by two stable moving fronts with symmetric velocities (see Fig. 4.11 a). This symmetry is broken for $b \neq 0$ and the pitchfork bifurcation splits into a stable branch and a saddle–node bifurcation. This is known as the Nonequilibrium–Ising–Bloch bifurcation (NIB).

For the asymmetric case no stationary front solution exists and the series expansion for small c cannot be applied. Alternatively we study the two–component system (4.12) for small values of . The series expansion reads

$$u(\ ) = u_0(\ ) + u_1(\ ) + {}^2 u_2(\ ) + (\ ^3),$$
$$v(\ ) = v_0(\ ) + v_1(\ ) + {}^2 v_2(\ ) + (\ ^3), \qquad (4.23)$$
$$c = c_0 + c_1 + {}^2 c_2 + (\ ^3).$$

Similar to the expansion for small velocities in the 0. order approximation the system collapses to an one–component reaction–diffusion dynamics:

$$Du_0'' + c_0 u_0' + u_0 - \frac{1}{3}u_0^3 - v_0 = 0, \quad c_0 v_0' = 0 \;\Rightarrow\; v_0 = \text{const.} \qquad (4.24)$$

comparable to Eq. 4.15 but with finite front velocity c_0. The activator solution reads

$$u_0(\ ) = u_1^s + \frac{u_3^s - u_1^s}{1 + \exp(\pm\ )} \quad \text{with} \quad = \frac{u_3^s - u_1^s}{\sqrt{6D}},$$
$$\text{and} \quad c_0 = \sqrt{\frac{D}{6}}(u_3^s + u_1^s - 2u_2^s). \qquad (4.25)$$

Here, $u_{1,2,3}^s$ denote the activator values of the fixed points of Eq. 4.24 given by the expression in Eqs. 4.3 with (v_0). The argument v_0 is not specified by Eq. 4.24 but it can determined by the corresponding inhibitor value of the local fixed points given in Eqs. 4.2.

The expression for c_0 is necessary for further analysis. However, the expression of the 0. order solution of the front is infeasible to determine higher order expressions in the same way as done in the previous paragraph. For further progress of the series expansion approach a piecewise linear approximation of the cubic branches of the activator nullcline is required. We are able to describe the front shape between the nullcline's extrema at $u = \pm 1$ via linearization according to Eqs. 4.3. The solution of the first equation in Eqs. 4.24 neglecting the cubic term can be expressed within

4 Bistable wave fronts interacting with boundaries

the interval $-1 \leq u_{\text{lin}} \leq 1$ as

$$u_{\text{lin}}(\xi) = A \exp\left(\frac{\tilde{c}_0}{2D}\xi\right) \sin(\kappa\xi) + v_0 \quad \text{within} \quad \xi_1(u_{\text{lin}} = -1) \leq \xi \leq \xi_2(u_{\text{lin}} = 1), \tag{4.26}$$

where $\kappa = \frac{\sqrt{4D - \tilde{c}_0^2}}{2D}$. In order to have a monotonic front shape the derivative of this expression must vanish at the limits of validity. That determines conditions for the coefficient A and the linearized velocity $\tilde{c}_0$. The latter can be explicitly written as

$$\tilde{c}_0 = \frac{2\sqrt{D} \log\left(\frac{1-v_0}{1+v_0}\right)}{\sqrt{\pi^2 + \log^2\left(\frac{1-v_0}{1+v_0}\right)}}. \tag{4.27}$$

Both expressions for the 0. order front velocity from Eq. 4.25 (c_0) and from the approximated Eq. 4.27 ($\tilde{c}_0$) depend on the inhibitor value v_0. It can take three different values, corresponding to the upper and lower local stable fixed point and the unstable fixed point in between. These three velocity values corresponds to a stable positive and a stable negative propagation direction and an unstable velocity branch. They can be read at the ordinate of Fig. 4.11 (b) as the limit $\epsilon \to 0$.

For the 1. order in ϵ we obtain

$$Du_1'' + c_0 u_1' + c_1 u_0' + u_1(1 - u_0^2) - v_1 = 0, \quad c_0 v_1' + g(u_0, v_0) = 0. \tag{4.28}$$

The second equation for the inhibitor can be integrated when the linear approximation for the front shape $u_0 \approx u_{\text{lin}}$ is used. Inserting the integral expression into the activator equation yields:

$$Du_1'' + \tilde{c}_0 u_1' + u_1 = v_1 - c_1 u_{\text{lin}}' \quad \text{with} \quad v_1(\xi) = \frac{1}{\tilde{c}_0} \int_{\xi_1}^{\xi} (v_0 - au_{\text{lin}}(\xi) - b) \, d\xi. \tag{4.29}$$

The homogeneous form of the remaining activator differential equation can be solved by choosing $u_1 = u_{\text{lin}}'$. This derivative is orthogonal to the inhomogeneity and we can proceed as already done to obtain Eq. 4.18. The orthogonality condition leads to an expression for c_1:

$$\int_{\xi_1}^{\xi_2} (v_1 - c_1 u_{\text{lin}}') u_{\text{lin}}' \, d\xi = 0 \Rightarrow c_1 = \frac{\int_{\xi_1}^{\xi_2} v_1 u_{\text{lin}}' \, d\xi}{\int_{\xi_1}^{\xi_2} (u_{\text{lin}}')^2 \, d\xi}. \tag{4.30}$$

In Fig. 4.11 the $c - \epsilon$ relation approximated as $c(\epsilon) \approx c_0 + c_1 \epsilon$ is depicted as dashed–dotted lines for the symmetric pitchfork– and the biased saddle–node–bifurcation. They mark the linear slopes for the stable velocity branches at a strong time scale separation between the activator and inhibitor variable and hence clearly deviate

4.4 Continuous bistable front

from the numerically obtained result when becomes too large. The asymmetry

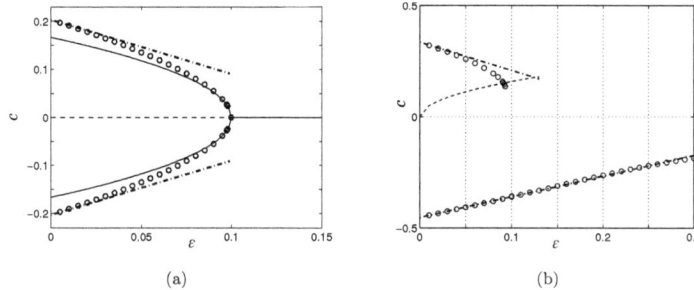

Figure 4.11: Characteristics of front velocity c in dependence on the time scale parameter from numerical simulations (symbols) and series expansions (lines).
(a): Pitchfork bifurcation for the symmetric case $b = 0$ ($a = 0.1$). Solid lines represent the stable branches of Eq. 4.20 from the series expansion in c. The linear approximation for small is shown as dashed–dotted lines. (b): Implied saddle–node bifurcation for $b = 0.05$ ($a = 0.2$). Again dashed–dotted lines represent the small approximation for stable solutions, the dashed line stands for the unstable solution of Eq. 4.33. Remaining parameter: $D = 1$.

due to $b \neq 0$ and the associated shift of the inhibitor nullcline in favor of one of the two stable fixed points results in a difference of the velocity values for both spatial directions. This is directly connected to the distances from the stable to the unstable fixed point. When for instance $u_3^s - u_2^s < u_1^s - u_2^s$, then the front prefers to cover the system with the u_3^s-state and propagates in the corresponding direction and vice versa. Nevertheless, for small there is always a front velocity directed to the the less preferred fixed point which is a small branch for positive velocities in case illustrated in Fig. 4.11. This branch annihilates in a saddle–node bifurcation with an unstable branch that also arrives from small . Unfortunately the exact location of the saddle–node bifurcation is not covered by a series expansion in up the linear order. However, by applying the –expansion for the unstable branch, it is possible to roughly estimate the position of the bifurcation point. The expansion

4 Bistable wave fronts interacting with boundaries

for the unstable branch has to be in orders of $\sqrt{\varepsilon}$:

$$u^u(\varepsilon) = u_0^u(\varepsilon) + \sqrt{\varepsilon}\, u_1^u(\varepsilon) + \varepsilon\, u_2^u(\varepsilon) + \mathcal{O}(\sqrt{\varepsilon}),$$
$$v^u(\varepsilon) = \sqrt{\varepsilon}\, v_1^u(\varepsilon) + \varepsilon\, v_2^u(\varepsilon) + \mathcal{O}(\sqrt{\varepsilon}),\qquad(4.31)$$
$$c^u = \sqrt{\varepsilon}\, c_1^u + \varepsilon\, c_2^u + \mathcal{O}(\sqrt{\varepsilon}).$$

Here, the index 'u' indicates the unstable solution. The same linearization ansatz as applied for the stable branches can be used and we arrive for the 0. order at

$$u_{\text{lin}}^u = 2\cos\left(\frac{\xi}{\sqrt{D}}\right) \quad \text{for} \quad -\sqrt{D} \leq \xi \leq \sqrt{D}. \qquad(4.32)$$

The **1. order** yields an orthogonality condition that yields a simple expression for c_1^u:

$$c_1^u = \sqrt{D(a+b)}, \qquad(4.33)$$

and thus an approximated relation for the unstable branch velocity for small ε, shown as dashed line in Fig. 4.11 (b). Obviously, in order to describe the curvature at the saddle–node value higher orders have to be considered. However, the negative branch, which is completely stable, follows the linear velocity dependence over a large ε range. The front propagation in space and time beyond but very close to

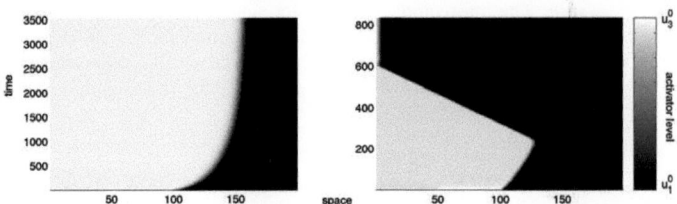

Figure 4.12: Space–time plots for two scenarios near bifurcations. The gray scale indicates the activator level. Left: close to pitchfork – $a = 0.1$, $b = 0$, $\varepsilon = 0.101$ and $D = 5.0$. Right: close to saddle–node – $a = 0.2$, $b = 0.05$, $\varepsilon = 0.0935$ and $D = 1.0$.

the pitchfork – and saddle–node bifurcation, is depicted in Fig. 4.12. For both cases initial conditions are chosen such that the front is driven to the right and thus to positive velocities. In the pitchfork case, neither the positive nor the negative branch exist for $\varepsilon > \varepsilon_{\text{crit}}$. So the initial velocity decreases and after a transient, the front stops moving. Beyond the saddle–node point there is one stable branch with a finite velocity left and thus, the front inverts its directions and returns to the left boundary. Interpreting the boundary interaction it is obvious that the front is not able to rebound due to the absence of a positive velocity branch.

The upper and lower stable velocity branches and their expansions $c(\varepsilon) = c_0 + \varepsilon c_1$ are also insightful in the representation over the asymmetry parameter b as shown in Fig. 4.13. The results from the series expansion are presented as solid lines that are in a good agreement with the numerical estimations (symbols) for small ε. Especially in the monostable area the linear approximation seems to be sufficient to estimate the velocities. We also notice that the bistable area is more extended the smaller ε is. Even stronger time scale separation would lead to an extension of the bistable velocity regime until values of b that are beyond the condition for a local bistable dynamics. The local system then undergoes a Hopf–bifurcation that terminates the stability of one of the stable fixed points. Also seen in the figure is the fact, that close to the bifurcation values, the evaluated linear expression fails and the critical points are missed.

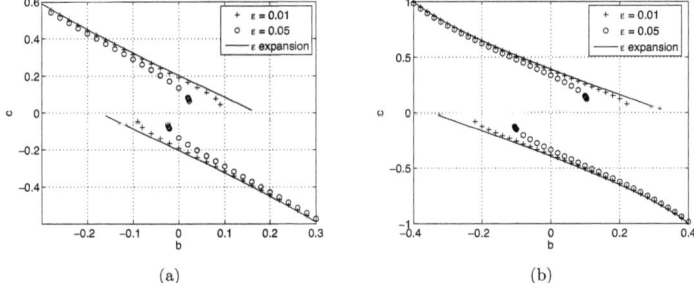

Figure 4.13: Bulk–velocities of the front versus b. Solid lines are due to series expansion in ε, symbols represent numerical simulations. (a): $a = 0.1$, (b): $a = 0.2$ and $D = 1$ for both.

4 Bistable wave fronts interacting with boundaries

4.4.2 Activator and inhibitor diffusion

If both, the activator and the inhibitor substance are mobile, the inhibitor is also described by a field $v(x,t)$ and its dynamics obeys a parabolic partial differential equation, too. The method presented here is similar to an approach of E. Meron et al. [46] that was however applied for a single–edge no–flux boundary.

The Eqs. 4.12 change into

$$\dot{u} = f(u) - v + D_u \Delta u \qquad (4.34)$$
$$\dot{v} = g(u,v) + D_v \Delta v.$$

The space can be rescaled by $x = \sqrt{D_u}\, y$, which leads to a inhibitor diffusion constant being the ratio $\delta = D_v/D_u$. The characteristic inhibitor diffusion length is then given by $l_D = \sqrt{\delta}$. For small δ this length may be remarkably greater than the activator diffusion range and the dynamics is dominated by the inhibitor diffusion. This system behaves rather different to the system analyzed in section 4.4.1 due to the non–vanishing inhibitor diffusion that governs the dynamics for $l_D \gg 0$. Within this parameter range, adiabatic elimination can be assumed to be valid, as $\delta \ll 1$. Therein the much faster activator process follows the inhibitor immediately and we can suppose that along the stable branches of the cubic nullcline $u(t) = u(v(t))$ holds.

In co–moving frame of the front coordinate is shifted to $\xi = y - ct$ transforming the partial differential equation for the inhibitor into a ordinary differential of second order.

$$\delta v''(\xi) + c v'(\xi) + \left(au(v) - v + b \right) = 0. \qquad (4.35)$$

For the inverted cubic branches $u(v)$ we apply the linearized expressions from Eqs. 4.4 resulting in two linear equations for the upper (positive) stable branch, indicated with '+' and for the lower (negative) branch indicated with '-' :

$$\delta v''_\pm(\xi) + 2v \gamma'_\pm(\xi) - \kappa \left(v_\pm - v^0_{3,1} \right) = 0, \qquad (4.36)$$

$$\gamma = \frac{c}{2}, \quad \kappa = -\left(a + \frac{1}{2}\right) \text{ and } v^0_{3,1} = \frac{2}{a+2}(b \pm \sqrt{3}a),$$

that have the solutions:

$$v_\pm(\xi) = e^{-\gamma \xi} \left(A_\pm e^{\lambda \xi} + B_\pm e^{-\lambda \xi} \right) + v^0_{3,1}, \qquad (4.37)$$

with $\lambda^2 = \gamma^2 + \kappa > 0$. To make the scenario comparable to the previous section we consider a wave shape that is fixed at the stable equilibrium values of the local dynamics v^0_3 at the left boundary and v^0_1 at the right boundary, as illustrated in Fig. 4.4. The space between the two boundaries has the length L so that the left

4.4 Continuous bistable front

handed boundary is located at $y = 0$ whereas the right handed boundary at $y = L$. Then the intermediate front has a certain distance to the left boundary, denoted as d. It is measured from the value of the unstable fixed point located on the inhibitory front

$$v(= d) = v_2^0 \qquad (4.38)$$

which is different from zero for $b \neq 0$.

The Dirichlet boundary conditions for the inhibitor field read $v_+(-d) = v_3^0$ and $v_-(L-d) = v_1^0$ which are used as initial conditions for Eqs. 4.36. They determine two of the four coefficients $A\pm$ and $B_\pm$. As a merging condition we assume that the upper and lower front parts meet at a certain front value v_F, where also differentiability is requested. That results in three further conditions: $v_+(0) = v_-(0) = v_F$ and $v'_+(0) = v'_-(0)$. With these five conditions not only the four coefficients can be fixed, in addition to it we have one equation to determine the distance d. Hence, the shape of the moving front would given as well as its position in between the two boundaries if the value of v_F, which remains unknown, could be fixed.

The determination of the inhibitor front value v_F combines the propagating activator front with its inhibitory counterpart. Assuming the variable v to be constant for a moment, the front velocity for the one–component reaction–diffusion system with a cubic nonlinearity is known for the case of natural boundary conditions. With the help of the series expansion we already found an expression for the velocity of a moving front given in Eq. 4.25. Here we treat the first equation of the set 4.34 with a rescaled diffusion constant $D_u = 1$. The activator fixed point values $u_{1,2,3}^s$ included in Eq. 4.16 depend on the specific value of v which we identify with the front value v_F in the actual case. Using the mentioned nullcline–linearization the activator bulk velocity can be written in a compact form:

$$c_a = -\sqrt{\frac{3}{2}} v_F + (v^2). \qquad (4.39)$$

We can assume that both the activator and the inhibitor front move with the same velocity. Then, Eq. 4.39 provides the remaining inhibitor front value. Even though this system acts in a different manner as in the situation of pure activator diffusion, the front motion far from the boundaries as well as close to them shows the same types of bifurcation behavior as discussed in the previous sections. As confirmed by numerical simulations of the Eqs. 4.34 we find again reflection or sticking at the boundaries, dependent on the choice of parameter values.

In the following we will describe the front motion in the phase space of its distance d to the boundary at $y = 0$ and the variation of the distance identifiable as front velocity $\dot{d} = c$. Trajectories in that phase space follow the dynamical equations:

$$\dot{d} = c, \quad \dot{c} = h(d, c). \qquad (4.40)$$

4 Bistable wave fronts interacting with boundaries

Within this picture we will obtain limit cycles that correspond to reflection on both boundaries and trajectories, that end in stable fixed points at the $c = 0$ nullcline, where the front stops moving. By extending the adiabatic elimination approach we will arrive at an expression to approximate the unknown function $h(d, c)$ and thus the nullcline dynamics for $\dot{c} = 0$.

First, we want to study trajectories close to the $\dot{d} = 0$ nullcline. Either a trajectory crosses this nullcline perpendicularly according to a front that changes the moving direction. Or a trajectory approaches the nullcline and ends in a stable point corresponding to a stationary front. These points are mostly located close to a boundary, where they can be described approximately by the decaying part of the front solutions of Eqs. 4.37 given as

$$\text{left: } v_-(y) = (v_3^0 - v_1^0)\, e^{-\sqrt{y}} + v_1^0,$$
$$\text{right: } v_+(y) = (v_1^0 - v_3^0)\, e^{-\sqrt{(L-y)}} + v_3^0. \qquad (4.41)$$

At the very edge of the system either the upper or the lower part of the wave description is sufficient in order to cover the front shape. Both cases are illustrated in Fig. 4.14 for a symmetric wave ($b = 0$) and a system length $L = 40$. The results from Eqs. 4.41 (dashed lines) are compared to numerical simulations, indicated as symbols in the figures. Shown is a situation, where the front is completely relaxed at the boundary holding its position. Due to Eq. 4.39 the front inhibitor value is zero and the distances to the corresponding boundaries are given by

$$\text{left: } d_0 = -\sqrt{\frac{2}{(a+2)}} \ln\left(\frac{v_1^0}{v_1^0 - v_3^0}\right),$$
$$\text{right: } d_0 = -\sqrt{\frac{2}{(a+2)}} \ln\left(\frac{v_3^0}{v_3^0 - v_1^0}\right). \qquad (4.42)$$

The points $(d_0, c = 0)$ at the left and at the right boundary define fixed point locations in the considered phase space and are marked as unfilled circles in the Figs. 4.16-4.18.

Beyond the estimation of these stationary points the Eqs. 4.37 serve the possibility to obtain an expression for a distance–velocity relation $h(d, c)$. By applying boundary and merging conditions and the inclusion of Eq. 4.39 we find explicitly

4.4 Continuous bistable front

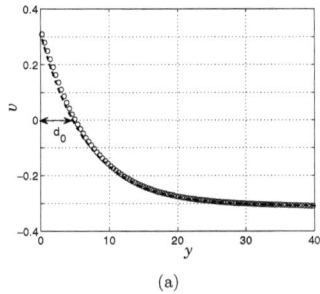
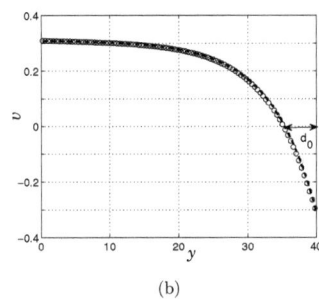

(a)   (b)

Figure 4.14: Inhibitor front shape at the left (a) and right (b) boundary follows exponential decay according to Eqs. 4.41 (dashed line) compared to simulations (symbols). Parameter values: $a = 0.2$, $b = 0$, $= 0.02$, $= 1$ and $L = 40$.

solved for the distance:

$$d(c) = \frac{1}{2\;(c)} \ln\left[\frac{1}{2\Delta(\;(c) - ^{(*)}\;(c))}\Big(\;(c)\Delta(1 + \;(c)) + \right.$$
$$\sqrt{4\;(c)\;\Delta^2 + (\;(c)\Delta(1 + \;(c)) + \;(c)\;(c)(\;(c) - 1))^2}$$
$$\left. + \;(c)\;(c)(\;(c) - 1)\Big)\right] \quad (4.43)$$

with $(c) = e^{2\;(c)L}$, $\Delta = v_1^0 - v_3^0$ and $(c) = v_1^0 + v_3^0 - 2v_F(c)$.

The star indicates the sign that is to change to a plus sign when no-flux boundaries are used instead of Dirichlet boundaries. Before we discuss this longish formula let us regard the asymptotic behavior. By shifting both boundaries to infinity, so that the front is always far away from having any boundary influence, we obtain the bulk velocity with its parameter dependencies. For presentability solved for b it reads

$$b = \frac{a+2}{4}\left(2v_F(c_\text{bulk}) + \frac{(c_\text{bulk})}{(c_\text{bulk})}(v_3^0 - v_1^0))\right). \quad (4.44)$$

This expression contains the regions of mono- and bi-stability and transitions via saddle-node bifurcations, shown over parameters b and in Fig. 4.15. The left figure (a) shows the positive and negative velocity branch (solid lines) which are symmetrical with respect to the sign of b. This type of velocity dependency is qualitatively the same as we have obtained in the previous section where a pure activator diffusion scenario was considered. For this system we found approximations

4 Bistable wave fronts interacting with boundaries

of stable velocity branches as shown in Fig. 4.11 and Fig. 4.13. For the current system with two diffusing variables we even can locate the unstable branches represented as dashed line in the figures. The stable velocities can be compared with numerical

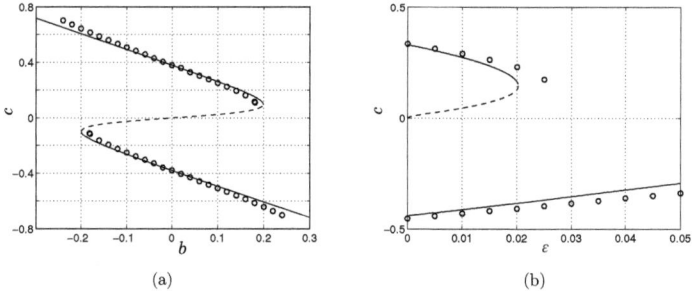

Figure 4.15: Bifurcation diagrams for bulk velocities. Solid and dashed lines are due to Eq. 4.44 and circles represent numerical simulations. (a): versus asymmetry parameter b for fixed = 0.01, = 0.1, (b): versus time scale separation parameter = 0.01 for fixed b = 0.05, = 0.5 and a = 0.2 for both.

simulated wave fronts (circles) by integrating the discretized RDS given by Eqs. 4.34. A sufficient number of elements were taken into account, so that boundary influence can be assumed to be negligible and the bulk velocity can be reached. Because of the assumption that the activator follows immediately the inhibitor dynamics it is clear that the smaller is, the better the theoretical curves match the numerical points. Thus the agreement in figure (a) is almost perfect for the chosen value of = 0.01 whereas in the figure with the dependence the differences become stronger with increasing time scale separation.

In Fig. 4.11 and Fig. 4.13 the same parameter values are used except the missing inhibitor term, corresponding to = 0. Comparing the diagrams showing the bulk velocity with respect to b, where in the inhibitor diffusion case = 0.1 is used the quantitative correspondence is obvious. Not only the bifurcation points are located both at $b \approx 0.2$, also the velocity values are similar. However, for = 0.5 in the diagram showing the dependence on the bifurcation values are clearly different, although the velocity value at which the upper branch looses stability is comparable.

In a final step of this chapter we will discuss examples of boundary interaction approximately described by Eq. 4.43 and studied by numerical simulations as well. Fig. 4.16 and Fig. 4.17 show the modes of bistable fronts interacting with two boundaries in the distance-velocity phase space. Depicted are solid lines corresponding to

4.4 Continuous bistable front

trajectories in (d, c) space which are obtained numerically. These trajectories represent the evolution of the position at the inhibitor front taken at the unstable fixed point value v_2^0 with its current velocity. The propagation direction is marked through attendant arrows on the trajectories. Also shown are dashed lines that present the results due to Eq. 4.43 and circles corresponding to fixed points given in Eqs. 4.42.

Two characteristic modes of movement and boundary contact behavior are to distinguish. Both can be found in Fig. 4.16 (a). Trajectories starting in the positive velocity region propagate a certain distance due to their initial configuration of the activator and inhibitor wave before they end in one of the both fixed points close to the boundaries. However there is also front propagation reaching the bulk velocity and covering a finite distance until it get attracted by a fixed point. Wether

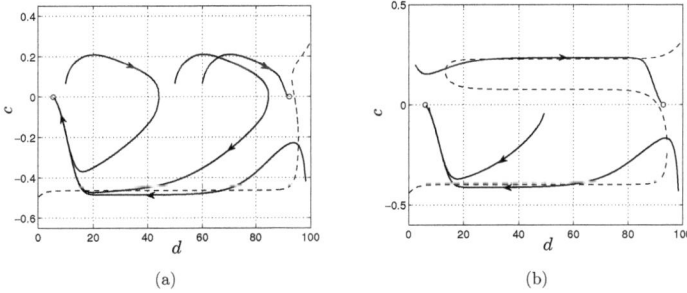

Figure 4.16: Trajectories as solid lines in distance–velocity (d, c) phase space and $d(c)$ relation from Eq. 4.43 (dashed line) are shown. Open circles represent the estimated location of fixed points along the $c = 0$ line. Both pictures illustrate cases with two stable points close to the boundaries where the front comes to rest. (a): For $b = 0.1$ the only stable direction is at $c < 0$ and the basin of attraction of the right handed fixed point is much smaller than that of the left handed fixed point. (b): For $b = 0.05$ an upper velocity branch for $c > 0$ appears maintaining the attractiveness of the right handed fixed point. Remaining parameter values: $a = 0.2$, $= 0.01$ and $= 1$.

trajectories reach the bulk velocity and stay on this c-level for a certain while, or the front slows down after an acceleration period in the beginning can be predicted by considering the relation from Eq. 4.43. Calculating $d(c)$ for the specific parameter values either one stable velocity branch at negative velocities appears, as in Fig. 4.16 (a) or two additional branches occur, an upper stable and an unstable branch in between growing from the right handed boundary when b is decreased. They can

87

4 Bistable wave fronts interacting with boundaries

be identified with the region of mono – and bi–stability of the bulk velocities in the bifurcation diagram over b, illustrated in Fig. 4.15. Therefore it is clear, that in the cases where b is chosen to be close to zero both stable branches exist.

Reflexion at the boundaries is shown for two examples in the Figs. 4.17. The trajectory of a front that rebounds at the left as well as at the right boundary forms a periodic orbit on the d, c–plane, shown in Fig. 4.17 (a). For that symmetric case ($b = 0$) the front proceeds running from one boundary to the other. In the very moment of the turn, the inhibitor front does not relax fully into the exponential shape. Thus, the estimation of the points $(d_0, 0)$ from Eq. 4.42 (see the circles in the figure) loses validity. The distance–velocity relation approximates the saturation velocities and one part of reflecting behavior. However, it fails at the open ends where the relation predicts a finite velocity directly at the boundaries.

Choosing a b slightly different from zero changes the situation, as shown in Fig. 4.17 (b). Due to the broken symmetry the positive velocity branch is at smaller values than the lower branch. Additionally, the front still rebounds at the right handed boundary but stops at left boundary. The basin of attraction of the corresponding fixed point spans the negative velocity region. However, for initial conditions close to that fixed point but set in the positive velocity region the trajectory runs first to the right boundary before coming back to the left handed fixed point. This behavior resembles typical characteristics of an excitable system.

The addition of Gaussian white noise to the activator dynamics at the same parameter set as used for the latter example leads to fluctuations around the stationary front position. Noise forces the phase point out of the basin of attraction erratically corresponding to a front that detaches from the left boundary. Hence, the front performs stochastically the whole cycle in phase space. Some of such noisy trajectories are shown in Fig. 4.18 with its equivalent in space–time (figure b). Similar to an excitable neuron driven by noise, randomly recurrent spikes appear that are robust on their excursion against further influence of noise. Even subthreshold excitations can be observed as small elongations from the fixed point.

4.5 Conclusions

The investigation of bistable fronts as a simplified approach to study excitable waves was the main aspect of this chapter. Our central focus laid on the interaction of fronts with Dirichlet boundaries and on the understanding of the observed front reflection or binding at the boundaries.

At first we considered an array consisting of FHN units with activator coupling that can be analyzed on the basis of local nullclines. For such discretized chains the effect of propagation failure as well as parameter regions of reflection and binding at the boundaries could be quantified in the limit of a thin front (see Fig. 4.8). Transitions from the binding to reflexion regime has been identified as Hopf–bifurcations

4.5 Conclusions

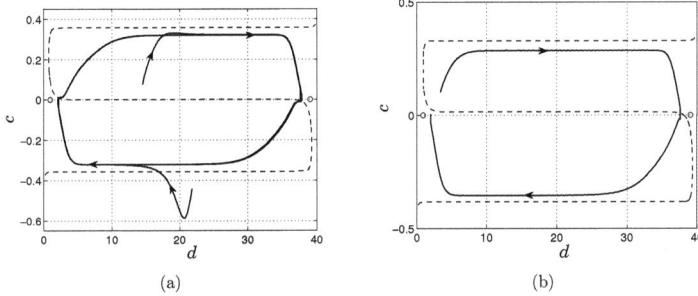

Figure 4.17: Phase space trajectories and $d(c)$ relation is depicted as in Fig. 4.16. (a): For that symmetric system ($b = 0$) both fixed points appear to be unstable and trajectories run into a periodic orbit indicating reflexion on both boundaries. (b): Slightly broken symmetry for $b = 0.02$. Left handed fixed point is stable where the trajectory ends after reflexion at the right boundary. Remaining parameter values: $a = 0.2$, = 0.05 and = 0.1.

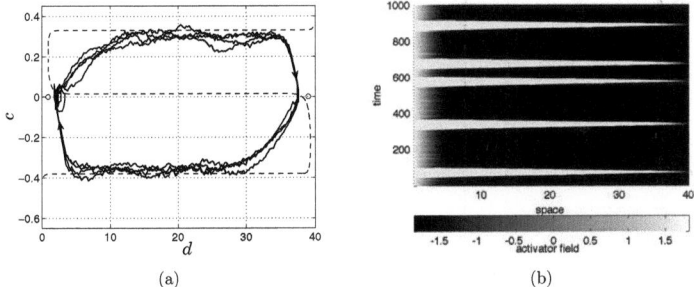

Figure 4.18: Dynamics under the influence of activator noise for the same parameter set as shown in Fig. 4.17. The system resembles the behavior of an excitable system including large excursions to the right boundary and back to the left handed fixed point. (a): phase portrait and (b) space time plot.

4 Bistable wave fronts interacting with boundaries

of the local stability. Close to these transitions units which are near boundaries and belonging to a stationary front profile exhibit oscillatory behavior with a small amplitude.

For a continuous bistable front we considered two cases; pure activator diffusion similar to the coupled discrete array and diffusion in both variables. The former has been analyzed without considering boundary effects. However, we found essential dependencies for the bulk velocity of the front by series expansion for small velocities as well as in the limit of strong time scale separation. Characteristic bifurcation scenarios such as saddle–node and NIB transitions have been estimated approximately. In addition we estimated the front shape and the corresponding trajectories in phase space.

When the inhibitor field diffuses in space accessorily while having a comparable diffusion coefficient then the system can be treated more extensive. For this purpose the cubic nullcline which defines the local activator dynamics had to be linearized. We found a relation between the front velocity and the distance of the front from the boundaries (Eq. 4.43) which gave us an estimative expression to quantify binding and reflexion of the front at the boundaries. Even the interaction with no–flux boundaries can be described by this expression. Furthermore, the limit of infinitely distant boundaries yields structurally the same bulk velocity behavior as found for the system with an immobile inhibitor.

To sum up we can fix, that seemingly complex interaction dynamics between a dissipative wave front and system boundaries as already mentioned in the chapter 3 can be understood and approximately quantified by analyzing bistable FHN–fronts.

5 Excitable two–state units coupled with delayed feedback

5.1 Introduction

The transition to oscillatory or bistable behavior from excitable dynamics was one of the foci in chapter 3 concerning the local FNK model. In such continuous systems those bifurcations emerge by tuning an appropriate control parameter which leads to an abrupt change of the dynamical behavior. They belong to the fundamental bifurcations in complex systems and are thus a well studied mechanism for single excitable units and for neuronal ensembles [3, 58, 149]. Within such ensembles the coupling type or its strength can cause those bifurcations although a single element exhibits excitable behavior, solely. This was exemplified in chapter 3 for the introduced FNK model where the chemical coupling via the external medium lowers the excitation threshold until a bifurcation occurs.

We will take up this idea in the present chapter by considering discrete states instead of continuous systems. The different dynamical states which are in the FHN system for instance the rest state, excited state and refractory state can be allocated to three discrete states where the transitions between the states are modeled by waiting time distributions [118]. Generally, the discrete state approach is a successful method applied to study various complex stochastic processes [94, 160, 128].

From Langevin equations describing the temporal evolution of continuous variables, we already discussed the transition to the Fokker–Planck equation. When the time scales of dynamical processes within the discrete states are short enough compared to the transitions between these states one can formulate a master equation as a type of a Fokker–Planck equation integrated over the basins of attraction in phase space, for which the discrete states are defined. Thus the master equations express the dynamics of probability to be in a certain state and their transitions to other states with probability currents.

In this chapter we introduce master equations for a single system possessing two states, for an ensemble of such systems coupled via instantaneous connections and with delayed coupling. The latter is caused by synaptic and dendritic signal transfer with finite velocity and introduces therefore an additional time scale that may lead to significant change of common dynamical behavior in ensembles of coupled neurons [47, 157, 19, 4, 23, 113].

Discrete models subjected to delayed feedback exhibit a great variety of dynam-

5 Excitable two-state units coupled with delayed feedback

ical features and they constitute a tractable way to control synchrony in neuronal, chemical or other individual based systems [106, 56, 70, 163, 146]. A model that consists of two states can be met in several applications, from studies of coherence and stochastic resonance to bifurcation analysis in networks of excitable or bistable units that exhibit oscillatory behavior [94, 106, 56, 70].

For the discrete states model investigated in the following the waiting times are given explicitly and bifurcations can be studied by comparing the time scales in a direct way. We analyze the statistical properties of an individual unit and the steady states of a globally coupled ensemble analytically and by numerical simulations. A bistable regime appearing due to saddle–node bifurcations can be found if the coupling includes a feedback that supports activity. When the feedback is delayed, an oscillatory regime emerges via a Hopf–bifurcation in the onset of coherent activation and synchronization of coupled units.

5.2 Individual unit

5.2.1 Model definition

The system under consideration, depicted in Figure 5.1 (a), constitutes a semi-Markovian process made up of two discrete states with specific waiting times distributed according to the density functions $w_1(t)$ and $w_2(t)$, respectively. So, the system spends a random amount of time in each state, however independent of the time spent in the preceding state. This single unit aims to mimic a single stochastic

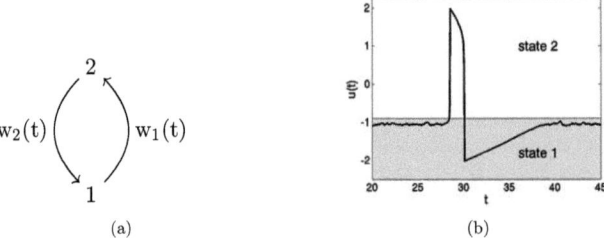

Figure 5.1: (a): Scheme of a two-state excitable unit. State 1 assigns as the resting state, followed by the excited state 2 with their waiting time distributions w_1 and w_2. (b): A typical spike of the FHN system in the excitable regime with areas of different gray levels denoting the states abstraction.

excitable system when specific distributions are used. In Fig. 5.1 (b) a typical spike event aroused by a noise induced excitation is shown in order to illustrate the cor-

5.2 Individual unit

respondency to the introduced states. The neuronal activation course (white area) belongs to state 2 while refractory and rest state are subsumed in state 1 (gray area). It resembles the same compartmentation as the step function did for the states in the FNK model controlling the release of potassium (see chapter 3.2.2). In terms of neuron models state 1 can be interpreted as the polarized state and state 2 as the the depolarized state, respectively.

It is evident that the whole dynamics of such a discrete process is fully specified by statistical properties of the waiting times between subsequent events. In our system the process is renewal, however the distributions of the waiting times are chosen to be unequal according to neuronal behavior. Thus the waiting times in the resting state obey different statistics than the waiting times in the excited state. The transition $1 \to 2$ is modeled as a rate process:

$$w_1(t) = \exp(-t), \qquad (5.1)$$

where both mean and variance equals $1/$. Eq. 5.1 describes the distribution of escape times needed, for an excitable system, to leave the rest state, under the influence of noise. The transition $2 \to 1$ is modeled by an Erlang distribution:

$$w_2(t) = \frac{a_2}{t_2 \Gamma(a_2)} \left(\frac{a_2 t}{t_2}\right)^{a_2-1} \exp\left(-\frac{a_2 t}{t_2}\right), \qquad (5.2)$$

where a_2 is integer. The mean value equals t_2 and the variance is given as t_2^2/a_2. This distribution describes the waiting time in the excited state corresponding to the quasi–deterministic spike production when the impact of fluctuations is negligible. Therefore the width of w_2 should be small, realized by choosing large values of a_2 where the Erlang distribution is close to a -distribution.

5.2.2 Interspike interval distribution and power spectral density

The renewal process describing the system above gives rise to a stochastic pulse sequence defined as $s(t) = 0$ in the rest state and $s(t) = 1$ in the excited state, illustrated in Fig. 5.2. Analyzing its statistical properties, characteristic features of excitable systems can be extracted.

The interspike interval (ISI) is the time between subsequent pulse events as shown in Fig. 5.2. The distribution of the ISI times can be expressed as the convolution of their waiting time density functions, since they are assumed to be independent. It is given as

$$\begin{aligned} w_{\text{ISI}}(t) &= \int_0^t dt'\, w_2(t') w_1(t-t') \\ &= \exp(-t) \left(\frac{a_2}{a_2 - t_2}\right)^{a_2} \left[1 - \frac{\Gamma(a_2, (a_2/t_2 -)t)}{\Gamma(a_2)}\right] \end{aligned} \qquad (5.3)$$

5 Excitable two–state units coupled with delayed feedback

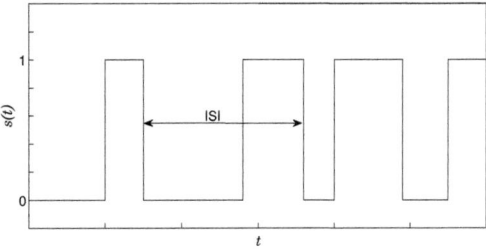

Figure 5.2: Rectangular shaped output sequence produced by one unit.

and depicted in Figure 5.3 (a). In the limit of a δ–distribution the ISI distribution is simply $w_{ISI}(t) = w_1(t)$.

The corresponding power spectral density (PSD) measures the power of each frequency ω occurring in the pulse sequence $s(t)$. It can be expressed in terms of the corresponding waiting time densities [136, 118]: (see Figure 5.3 (b))

$$S(\omega) = \frac{4}{\omega^2(t_2 + 1/\gamma)} Re \frac{[(1 - i\omega t_2/a_2)^{a_2} - 1]}{(\gamma + i\omega)(1 - i\omega t_2/a_2)^{a_2} - i\gamma}.\qquad(5.4)$$

When $1/\gamma \gg t_2$ (dashed lines in Figure 5.3) the unit spends more time in the resting state and the ISI distribution is broad. That corresponds to low noise intensity and, hence, rare pulse events. This is reflected in the fact that low frequencies collect more power. The PSD even has its highest maximum at $\omega = 0$.

Otherwise, if $1/\gamma \ll t_2$ (solid lines in Figure 5.3) the unit behaves like an oscillator between the two states with a spectral maximum at finite non–zero frequencies. The corresponding ISI are narrow distributed around the mean period of one cycle ($T = 1/\gamma + t_2$). Both PSD and ISI give evidence to coherence resonance, as there are some preferable frequencies that increase the pulse regularity [110, 38, 88, 119].

The choice of functions for the waiting time density distributions w_1 and w_2 seems to be plausible regarding the outcome for the ISI distribution and power spectral density. In the phase space picture of dynamical systems an increase of the activation rate γ can be interpreted as a lower excitation threshold around fixed point location. In case of FHN or FNK models a lower threshold gives evidence to a near Hopf–bifurcation point, which leads to a more resonant behavior of the system. In the section 3.3 both statistical measures are studied for the FNK model which is depicted in Fig. 3.7. To remind the reader, increasing the coupling C to the exterior in that model let the imaginary part of the eigenvalue belonging to the rest state grow, which led to an elevated excitation probability. Thus, the activation rate γ in the

5.2 Individual unit

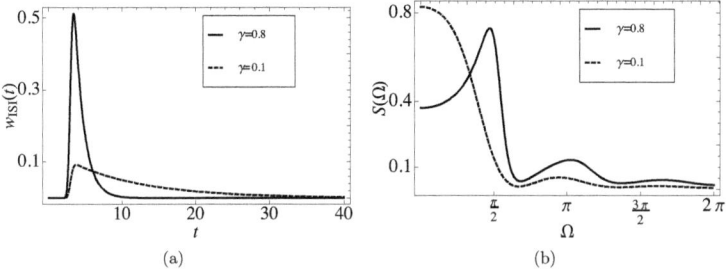

Figure 5.3: (a): Interspike interval distribution of a single unit with an exponential (w_1) and an Erlang (w_2) waiting time distribution. (b): Corresponding power spectral density for two different activation rates. Parameters used for Erlang distributions are $a_2 = 100$, $t_2 = 3$. The solid lines show an oscillatory unit, while the dashed lines display excitability.

two–state model corresponds to the coupling C in the FNK model within a certain range. A direct comparison of the ISI distribution and the power spectral density of the two–state model depicted Fig. 5.3 and for the FNK model shown in Fig. 3.7 shows the same change in their shape with increasing or C, respectively. This corroborates our assumption that the discrete states approximation is reasonable at least on the base of acting time scales.

5.2.3 Generalized master equations

The balance of probability flows serves to determine the occupation probabilities $P_i(t)$, $i = 1, 2$ of separate states i. The generalized master equations [69] that hold for these probabilities read:

$$\frac{d}{dt}P_1(t) = J_{2\to 1}(t) - J_{1\to 2}(t) \qquad (5.5a)$$

$$\frac{d}{dt}P_2(t) = J_{1\to 2}(t) - J_{2\to 1}(t), \qquad (5.5b)$$

where $J_{1\to 2}(t)$ and $J_{2\to 1}(t)$ denote the probability flow from state 1 to 2 at time t and vice versa. Since the transition $1 \to 2$ is a rate process, its probability flow is given by $J_{1\to 2}(t) = P_1(t)$. The second probability flow is given by the product of the inflow in state 2 in the past $J_{1\to 2}(t - t')$ with the waiting time density, $w_2(t')$ to

5 Excitable two–state units coupled with delayed feedback

wait the time t' in state 2, integrated for all possible t', namely

$$J_{2 \to 1}(t) = \int_0^\infty P_1(t-t')w_2(t')\,dt'. \tag{5.6}$$

Inserting the probability flows in Eq. (5.5b) yields:

$$\frac{d}{dt}P_2(t) = P_1(t) - \int_0^\infty P_1(t-t')w_2(t')\,dt'. \tag{5.7}$$

The equation (5.7) supplemented by the normalization condition $P_1(t) = 1 - P_2(t)$ can be given in the closed form:

$$\frac{d}{dt}P_2(t) = [1 - P_2(t)] - \int_0^\infty [1 - P_2(t-t')]w_2(t')\,dt'. \tag{5.8}$$

This integro–differential Eq. (5.8) characterizes the whole dynamics of a single unit.

It has to be supplemented by initial conditions obeying normalization and contains all the dynamical features of a single unit. So far the only parameter controlling the system is essentially the rate . While the waiting time distribution to be in state 2 is set to be close to a –distribution, this rate controls the distribution of occupation probability over both states.

5.3 Ensembles of globally coupled units

5.3.1 Instantaneous coupling

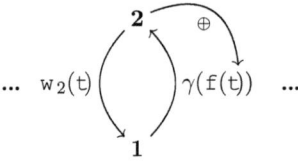

Figure 5.4: Scheme of a two–states unit coupled by the ensemble's global output $f(t)$.

The collective behavior of a large population of interconnected two–states units, illustrated in Figure 5.4, results from the interplay of individual participants. In order to investigate its dynamics, an ensemble of N units coupled by their global output $f(t)$, is considered. The information transmitted to each unit of the ensemble is the fraction of units in the excited state 2. Although the characteristics of the waiting time densities do not change due to coupling, the activation rate depends

5.3 Ensembles of globally coupled units

on $f(t)$. This mechanism can be expressed by the sum of output of all units

$$f(t) = \frac{1}{N}\sum_{j=1}^{N} s_j(t) = \frac{n_2(t)}{N}, \qquad (5.9)$$

where $n_2(t)$ is the number of units in the excited state at time t. Obviously, $f(t)$ equals the relative occupation number of the excited state. In the continuum limit the output is the mean occupation probability of state 2

$$\lim_{N\to\infty} f(t) = P_2(t). \qquad (5.10)$$

We thus consider the activation rate as a function of $P_2(t)$,

$$\Gamma = \Gamma(P_2(t)), \qquad (5.11)$$

and as the same for all the units. The equation that governs the collective dynamics of the ensemble is uniquely determined by the equation (5.8) for occupation probabilities of individual units. To that end, by taking Eq. (5.11) into account, the Mean–Field (MF) equation reads:

$$\frac{d}{dt}P_2(t) = \Gamma(P_2(t))[1 - P_2(t)]$$
$$- \int_0^\infty \Gamma(P_2(t-t'))[1 - P_2(t-t')]w_2(t')\,dt', \qquad (5.12)$$

in which the transition rate from state 1 to 2 is now implicitly time dependent. In general, equation (5.11) induces an implicit form of steady state solutions, which can be expressed as

$$P_1^* = \frac{1/\Gamma(P_2^*)}{1/\Gamma(P_2^*) + t_2}, \qquad P_2^* = 1 - P_1^* = \frac{t_2}{1/\Gamma(P_2^*) + t_2}. \qquad (5.13)$$

The reader is advised to consult the appendix B for calculation details. In Eq. 5.13 can be seen that the stationary probability is given as the fraction of mean waiting times within the states independently on the width of w_2 controlled by a_2. Note that units belonging to the stationary system still pass through the two states following their individual waiting times. However, regarding the whole ensemble, the fraction of units being in state 1 and state 2 remain constant. Stability of the steady states given in Eq. 5.13 can be acquired by adding small perturbations to the stationary probabilities: $P_i(t) = P_i^* + \epsilon_i \exp(t) + O(\epsilon_i^2)$ ($i = 1, 2$). Thus the rate expands up to the first order in the perturbations as

$$\Gamma(P_2) = \Gamma(P_2^*) + \epsilon_2 e^t \frac{d}{dP_2}\Gamma(P_2)|_{P_2^*} + O(\epsilon_2^2). \qquad (5.14)$$

5 Excitable two–state units coupled with delayed feedback

Inserting both perturbed quantities into the MF–equation 5.12 leads to the characteristic equation for the eigenvalues. It can be formulated as:

$$+ [\,\alpha(P_2^*) - (1 - P_2^*)\frac{d}{dP_2}\alpha(P_2)|_{P_2^*}][1 - \hat{w}_2(\,)] = 0, \qquad (5.15)$$

where $\hat{w}_2(\,)$ is the Laplace transform of $w_2(t)$:

$$\hat{w}_2(\,) = \mathcal{L}[w_2](\,) = \int_0^\infty dt\, w_2(t) e^{-t}\,. \qquad (5.16)$$

In case of no influence of fluctuation on the excited state the waiting time is δ–distributed around its mean t_2 and its Laplace transform is simply: $\hat{w}_2(\,) = e^{-t_2}$. Then the characteristic equation takes the form:

$$+ \Gamma[1 - e^{-t_2}] = 0, \qquad (5.17)$$

where Γ stands for the first expression in brackets in Eq. 5.15. The eigenvalue can be separated in real and imaginary parts: $\,=\,\mathrm{Re}\,+i\,\mathrm{Im}\,=\,+i\Omega$. Searching for a parameter region where $\mathrm{Re}\,=0$ leads to:

$$\mathrm{Re}:\ \Gamma(1 - \cos\Omega t_2) = 0\ \text{ and }\ \mathrm{Im}:\ \Omega + \Gamma\sin\Omega t_2 = 0\,. \qquad (5.18)$$

For any $\Omega \neq 0$ the first condition requires $\Gamma = 0$ and thus the second condition yields $\Omega = 0$. This contradiction forbids vanishing real parts and hence, oscillatory behavior of the ensemble due to the introduced feedback is excluded. The same can be shown for any waiting time distribution w_2 [117]. Therefore no Hopf–bifurcation appears through a coupling in the activation rates that would lead to oscillations between the two states.

However, an appropriate choice for the dependence of α on $P_2(t)$ can induce saddle–node bifurcations and bistability into the ensemble dynamics determined by Eq. (5.12). In the following we discuss a specific dependency of the rate on $P_2(t)$ from which we require a direct feedback to the activation rate. That means we construct an expression that leads to either positive or excitatory feebback, which heightens the activation rate the more units spend their time in state 2. Or, vice versa, a negative or inhibitory feedback can be considered, for which units occupying state 2 lower the activation rate.

We will quantify bifurcations for an Arrhenius law–like activation rate α in particular according to a potential barrier concept. This is motivated by the noise induced escape process from the rest state 1 interpreted as diffusion over a barrier (see chapter 2.2). Inhibitory as well as excitatory coupling can be modeled using

$$\alpha(P_2(t)) = \alpha_0 \exp\left(-\frac{\Delta U_0}{}(1 \mp P_2(t))\right), \qquad (5.19)$$

5.3 Ensembles of globally coupled units

where the latter requires the '−' sign in the last term. Here ΔU_0 is a potential barrier, is the noise intensity and is the coupling strength. This adoption fulfills the conditions for Kramers' time $(1/\ )$ when is sufficiently low. It has been used in studies of coherent and stochastic resonance [88, 110] and in globally coupled networks of bistable elements [55, 62, 118].

We will focus on excitatory feedback, a type of mutual influence that we already obtained in the response of the FNK model. We will briefly recall that mechanism for the reader. In the FNK model active units deliver potassium into the extracellular space that lowers the activation threshold in turn. That constitutes a typical positive feedback dynamics. Using the excitatory coupling due to Eq. 5.19 and inserting this expression into Eq. 5.15 we can study the stability of stable fixed points quantitatively.

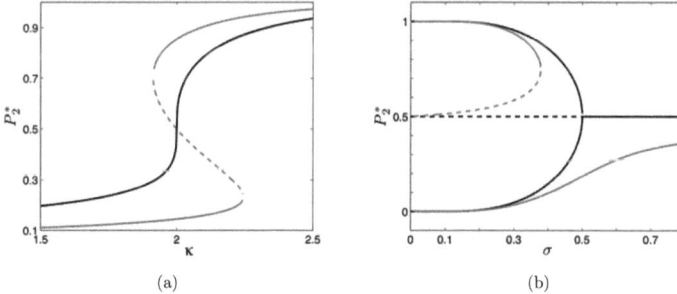

Figure 5.5: Bifurcation diagrams of the stationary steady state P_2^*. (a): Steady states with respect to coupling strength, monostability: = 0.5 as black line (pitchfork point), bistability: = 0.4 as gray lines, $r_0 = 1$ for both. (b): Steady states with respect to noise intensity, pitchfork at $r_0 = 1$ (black lines), saddle–node at $r_0 = 0.8$ (gray lines). Remaining parameter values: $t_2 = 1$, = 2, $\Delta U_0 = 1$.

For low coupling values the ensemble consisting of infinitely many units possesses one stable stationary configuration wherein the majority of units are in the rest state, as shown in Figure 5.5 (a). Following this stable branch for intermediate , a bistable regime appears due to a saddle–node bifurcation, where an additional stable and unstable fixed point at a higher P_2 value is created (gray curve in Figure 5.5 (a)). The unstable branch annihilates with the stable lower fixed point at higher coupling values through a second saddle–node bifurcation and the upper stable fixed point remains. The system leaves bistability and only the upper branch survives, where the majority of units are excited. Bistability appears for noise intensities below a

5 Excitable two–state units coupled with delayed feedback

critical noise value ^{crit} at which a pitchfork bifurcation occurs. This is illustrated as a black line in both panels of Fig. 5.5.

The critical pitchfork values can be determined by comparing the curvature of the expressions for the steady states in Eqs. 5.13. Applying the second derivative on the fixed point expression for P_2 and using the Arrhenius–type rate from Eq. 5.19 we obtain

$$t_2 = \frac{1}{(P_2^*)} \quad \text{and} \quad ^{\text{crit}} = \frac{\Delta U_0}{4}. \qquad (5.20)$$

The first identity reflects the merging of time scales, where the mean waiting times in both states are equal and the second equation provides the noise value at the pitchfork bifurcation point for a given coupling. The whole scenario is illustrated in the (–) parameter diagram Figure 5.6 where the light gray region stands for the bistable regime whereas the dark gray region depicts monostability. The border between them shows the parameter values where the saddle–node bifurcations take place merging in a cusp point collapsing to the pitchfork bifurcation. Note that the same bifurcation pattern was found for moving reaction–diffusion fronts, discussed in chapter 4. The insets in Fig. 5.6 show the corresponding left and right hand sides of the fixed indicate Eq. 5.13 for P_2^* whose number of cross sections point the monostable or bistable behavior. It is important to note that although the upper branch is stable, a single unit leaves the ensembles' steady state after its individual excitation time. However the ensemble reaches one of two states, determined by the system parameters and the initial configuration of units distributed in both states.

5.3.2 Delayed coupling

Up to this point the global output feeds back immediately to the rate . However, due to finite propagating velocity of signals in neural networks we assume a feedback that needs a certain but fixed delay time to act on the individual activation times. That addition to the model goes without any geometrical assumption or, in other words, every participating unit is assumed to have the same distance to any other unit. Therefore there is no heterogeneity in the feedback delay coming from the entire system. The introduction of such a delayed feedback induces significant variations on the ensemble dynamics. In the same manner as before, the rate is assumed to be an increasing function of $P_2(t -)$, meaning that the activation rate depends on the fraction of units that were excited at a certain time in the past. Extending Eq. 5.12 by the delay the MF–equation reads:

$$\frac{d}{dt} P_2(t) = (P_2(t -))[1 - P_2(t)]$$
$$- \int_0^\infty (P_2(t - t' -))[1 - P_2(t - t')] w_2(t') \, dt'. \qquad (5.21)$$

5.3 Ensembles of globally coupled units

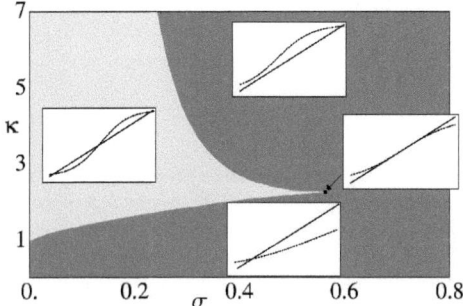

Figure 5.6: Stability diagram on — plane. The monostable regime is shown in dark gray, while light gray corresponds to the bistable regime. The borderline defines the positions of saddle–node bifurcations, merging in the cusp point of the pitchfork bifurcation (black dot). The insets in the corresponding regimes show geometrically the stationary states P_2^* from Eq. (5.13). The remaining parameters are the same as in Figure 5.5.

It is clear that the stationary states in the delayed system are the same as in the case of instantaneous coupling, given in Eqs. 5.13. However, the stability or the location of bifurcations can be different. Besides saddle–node bifurcations even a Hopf–bifurcation can be induced due to the introduction of delayed feedback [43]. It is characterized by vanishing real part of the fixed point's eigenvalue simultaneously with a finite imaginary part, determining the oscillation frequency directly at the bifurcation.

Linear stability analysis leads to the appearance of a factor $\exp(-\ )$ in the linearized rate:

$$(P_2) = (P_2^*) + {}_2 e^{(t-\ )}\frac{d}{dP_2}(P_2)|_{P_2^*} + (\tfrac{2}{2}), \qquad (5.22)$$

which propagates into the characteristic equation:

$$+ [\ (P_2^*) - (1 - P_2^*)\exp(-\ )\frac{d}{dP_2}(P_2)|_{P_2^*}][1 - \hat{w}_2(\ )] = 0. \qquad (5.23)$$

Oppositely to the previous section for a given set of parameters Eq. 5.23 has complex solutions that pass the imaginary axis simultaneously.

In the following, deviations in the excitation time are completely excluded. By setting $a_2 \to \infty$ the transition $2 \to 1$ has a fixed time t_2 and the waiting time in

5 Excitable two–state units coupled with delayed feedback

state 2 is -distributed, namely $w_2(t) = (t - t_2)$. As already mentioned for the limiting case in Eq. 5.17, its Laplace transform is $\hat{w}_2(\ ) = \exp(t\ _2)$ and thus the characteristic equation, which we will examine, reads:

$$+ [\ (P_2^*) - \frac{d\ (P_2^*)}{dP_2^*}(1 - P_2^*)\exp(-\ )][1 - \exp(t\ _2)] = 0. \qquad (5.24)$$

Now, three time scales struggle for dominance, these are first the activation time $1/\ $, second the residence time t_2 in state 2 and finally the delay time of the feedback $\ $. In order to obtain the parameter values at which the Hopf–bifurcation takes place, we separate Eq. (5.24) into real and imaginary part and set $\ =\ +\imath\ $. For $\ = 0$ the characteristic equation yields two independent expression in which we substitute the rate with $\ = P_2^*/(t_2 P_1^*)$, taken from Eq. 5.13 for the stationary states. By setting the real and imaginary part equal zero we obtain:

$$\frac{P_2^*}{1 - P_2^*} = -2t\ _2\left(\cot(\ ) + \cot(\frac{t\ _2}{2})\right) =: h(\ ) \qquad (5.25a)$$

$$\text{and} \qquad _{\text{Hopf}} = -\frac{t\ _2}{2\Delta U_0 P_2^*\sin(\ )}. \qquad (5.25b)$$

Inverting the function $h(\ )$ provides an expression for the critical oscillation frequencies $\ $ at the Hopf–bifurcation depending on parameter values implicated in P_2^*, shown in the inset of Figure 5.7 over noise intensity. The second equation is an implicit expression for the critical parameters on the $\ -\ $ plane where the transition to coherent oscillations takes place. The parametric curve is shown for different delays in Figure 5.7. The frequency, needed in this expression, can be taken from Eq. 5.25a.

The location and shape of the curve changes with different time delay and divides the $\ -\ $ plane in two dynamical regimes, an oscillating and a non–oscillating domain. Inside the oscillating regime all units undergo the transitions $1 \to 2 \to 1$ in a coherent way leading to an oscillatory global output. Outside this area the ensemble ends up in a stationary state and exhibits no rhythmic phenomena, although all units change between both states following their individual waiting time distributions. The critical oscillation frequency $\ $ along the bifurcation curve increases monotonically with larger noise intensity, as shown in the inset of Figure 5.7. The area with oscillations in the $\ -\ $ plane shrinks when the delay decreases. As one limiting case we know from the previous section that no oscillatory regime exists for $\ = 0$. There is a minimal but non–zero delay below which no Hopf–bifurcation emerges. An analytical approximation of this critical delay can be estimated considering Eq. 5.25b. With respect to $\ $ it takes the form $\sin(\ ) = -\ $, where $\ $ contains the remaining parameters. The parameter range we are considering makes

5.3 Ensembles of globally coupled units

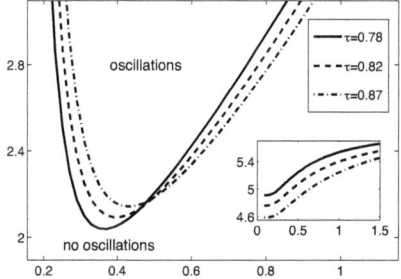

Figure 5.7: Curves showing the appearance of Hopf–bifurcations in the parametric — plane for different coupling delays. The associated frequencies along the Hopf curves are shown in the inset. Remaining parameters are fixed at $t_2 = 1$, $r_0 = 0.8$, $\Delta U_0 = 1$ and $a_2 \to \infty$.

positive and smaller than one and thus solutions of the transcendental equation are possible for certain values of . In fact, they can be approximated geometrically, assuming that the linear term touches the sinus close to its first minimum at $= 3/2$. Using this estimation for Eq. 5.25b gives

$$\text{critical} \approx \frac{3}{4} \frac{t_2}{\Delta U_0 P_2^*}. \qquad (5.26)$$

This critical value defines the position of the primary branch in Figure 5.8. From the parametric equations (5.25) for the Hopf–bifurcation points, additional solutions can be easily derived, given by,

$$' = + \frac{k}{}, \qquad (5.27)$$

where $k = 0, 1, 2....$ Equation (5.27) indicates reappearance of Hopf–bifurcations and periodic solutions for larger delays. The primary branch of these solutions appears at finite delays for $k = 0$ as shown in Figure 5.8. The reappearance of this primary branch for larger delays ($k = 2, 4, ...$) is also shown in the same figure. These branches have the same frequency dependence, since they consist of the same periodic solutions (for even values of k), however with some stretching and squeezing. Let us assume two solutions with different frequencies for delays $_1$ and $_2$, with $_1 < _2$, on the primary branch. The projection of their distance on the –axis is

5 Excitable two–state units coupled with delayed feedback

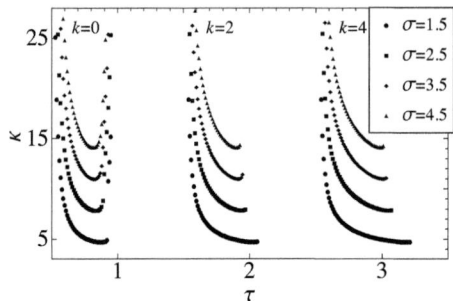

Figure 5.8: Reappearance of the primary Hopf branch ($k = 0$) for longer delay times shown for different noise intensities in an ascending order from small to high noise. Some parts are squeezed and some parts are stretched, depending on their frequency as estimated in Eq. 5.29. Remaining parameters are the same as in Figure 5.7 .

defined as:

$$\triangle = \tau_2 - \tau_1 . \quad (5.28)$$

The distance of the same solutions on the k-th branch is given by using Eq. (5.27) as,

$$\triangle' = \tau'_2 - \tau'_1 = \triangle + k \frac{\omega_1 - \omega_2}{\omega_1 \omega_2} . \quad (5.29)$$

From equation (5.29) it is clear that $\triangle' > \triangle$ when $\omega_1 > \omega_2$ and thus the corresponding parts of k-th branch will be stretched, otherwise they will be squeezed. This seems to be a general behavior of a delayed system with periodic solutions [162].

Finally, we want to point out that from Eq. (5.25b) follows that also for negative coupling, values $\kappa < 0$, corresponding to inhibitory coupling the Hopf–bifurcation line exists. Namely, due to Eq. (5.27) odd values of k lead to a $-\sin(\omega \tau')$ term entering in Eq. 5.25b which inverts the sign of κ_{Hopf}. So, feedback acting delayed with τ'_k and odd k induces coherent oscillations in the system for inhibitory coupling. However, the bifurcation line has not the same shape as in Fig. 5.7 mirrored at the κ axis, but the transition occurs at different values of noise intensity. That becomes clear keeping in mind the implicit dependency on κ and σ on P_2^*. The different sign of inhibitory coupling shifts the stationary states and therefore also the bifurcation lines.

5.4 Numerical simulations

The dynamics of a globally coupled ensemble constituted of a large number of two–states units is simulated in order to verify the theoretical results presented above. In the absence of delayed feedback and with an Erlang distribution for the waiting times in the excited state 2, monostable and bistable behavior is observed. In the bistable case one of the stationary states can be selected by choosing appropriate initial configurations. Note that the individual units behave in a different way along the two stable branches, as already mentioned for the infinite system at the MF level. For stationary occupation probabilities belonging to the lower stable branch the activation time can be very long. Then the majority of units stay in the resting state at almost all times only interrupted by rare transitions into state 2. Units belonging to the stationary state at the upper stable branch behave differently. Each unit leaves the excited state after a finite excitation time, however gets excited again after a vanishing short activation time. This complex behavior arises because the single elements always change between two states following their individual internal clocks while the lower stationary state is dominated by the time scale of activation, whereas the excitation time t_2 governs the upper stationary state.

By applying the delayed feedback and assuming a constant excitation time ($w_2(t) = (t - t_2)$), an additional time scale is introduced that can cause bulk oscillations. As predicted in the previous section, oscillatory behavior occurs for parameters and beyond the Hopf line given by Eqs. (5.25). In the figures 5.9 and 5.10 raster plots show the results of numerical simulations. The first figure contains two plots for fixed and different , thus crossing the bifurcation curve in Fig. 5.7 vertically. The first plot shows the case for a coupling value outside the oscillatory region and the second plot a case of collective oscillations. Fig. 5.10 contains three plots for different noise values and a fixed coupling, corresponding to a horizontal crossing of the bifurcation curve. Only the middle plot represents the oscillatory regime, the upper and lower plot stand for the left and right hand sides outside that region. In the upper panel of each figure the activity of 2500 delayed–coupled units is recorded, where the black dots mark the transition events to the excited state. The activity of an arbitrarily chosen individual unit (dashed line) and the global output (solid line) are depicted in the lower panel.

The situation for $= 0.4$ and $= 2$ depicted in Fig. 5.9 (a) is essentially the stationary solution, which we already have seen in the bifurcation diagram in Fig. 5.5 (a), where the global output is around 0.1, corresponding to the value of P_2^* in the diagram. Firing events are comparatively rare and irregular and no pattern in the raster plot emerges. In Fig. 5.9 (b) the system is shown for $= 2.3$ and hence beyond the Hopf–bifurcation line. The global output shows distinct oscillations evoked by the coherent transition of the individual units between states as it can be seen in the corresponding raster plot.

In Fig. 5.10 the coupling value is fixed for each of the three sub–figures at $= 2.3$.

5 Excitable two–state units coupled with delayed feedback

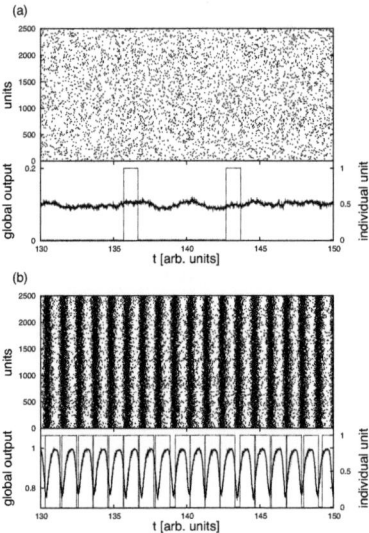

Figure 5.9: Raster plots (upper panels) and global output in addition with an arbitrarily chosen single unit (lower panels, solid and dashed lines) are shown as a record of the activity of 2500 units with delayed global coupling. Black dots stand for the moments where the transition $1 \rightarrow 2$ takes place. Delay value chosen for this simulation: $= 0.78$
(a): Non–oscillatory regime at $= 2$ and $= 0.4$,
(b): oscillatory regime at $= 2.3$ and $= 0.4$.

Starting with a noise intensity of $= 0.2$ we find a similar situation as stated in Fig. 5.9 (a), where the system occupies the lower stationary state involving single units which exhibit rare firing events. For $= 0.4$ the units perform synchronized transitions within a small time interval making the global output oscillate (Fig. 5.9 b). The oscillation period corresponds roughly to the residence time t_2 due to a very fast activation from state 1. Even higher noise lets the system cross the second Hopf–bifurcation and the oscillations disappear, as shown in Fig. 5.10 (c) for $= 0.6$. The firing events are more frequent, however, not synchronized and thus the common output remains constant apart from finite size fluctuations.

Two measures are applied to quantify coherent global oscillations. The very presence of oscillations can be captured by regarding the amplitude in the global output.

5.4 Numerical simulations

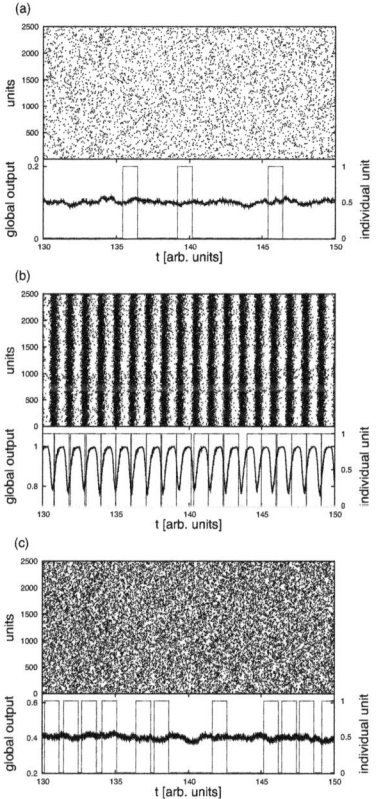

Figure 5.10: Raster plots and output record as in Fig. 5.9 for the same delay time.
(a): Non-oscillatory regime at = 0.2 and = 2.3,
(b): oscillatory regime at = 0.4 and = 2.3,
(c): non-oscillatory regime at = 0.6 and = 2.3.

Using the mean deviation from its average we define:

$$\text{amplitude} = \sqrt{\langle (f(t) - \langle f(t) \rangle)^2 \rangle}. \qquad (5.30)$$

5 Excitable two–state units coupled with delayed feedback

Outside the oscillatory region this quantity takes non–zero but negligible values due to the finite number of elements in the simulated ensemble. As soon as the Hopf line is passed the amplitude rises smoothly over , reaching a maximum and goes down to a value close to zero when the second Hopf–bifurcation is passed (Fig. 5.11 a). Over the amplitude rises abruptly to a maximum beyond the bifurcation point. At higher coupling strength the amplitude decreases, although there is no further Hopf–bifurcation. This is due to further separation of time scales. With increased coupling comes a higher activation rate and a short stay in state 1. Thus most of the units are in state 2 within a certain time interval and the coherent transition of a considerable number of units between the states, which causes elongations in the global output, becomes improbable.

The degree of coherence can be characterized by the synchronization index (S.I.). Following the analytic signal approach [111] of a measured signal $z(t)$, which is the global output in our case, one can define:

$$(t) = z(t) + i\tilde{z}(t) = A(t)e^{i\ (t)}, \qquad (5.31)$$

where the imaginary part $\tilde{z}(t)$ is the Hilbert transform of $z(t)$. The instantaneous phase (t) of the signal is uniquely defined by Eq. (5.31). Assuming two arbitrarily chosen subensembles, each consisting of 50 units, the phase difference between them is defined as, $\Delta\ =\ _1(t) -\ _2(t)$. Therefrom the S.I. can be estimated, given by

$$\text{S.I.} = \langle \cos(\Delta\ ) \rangle^2 + \langle \sin(\Delta\ ) \rangle^2. \qquad (5.32)$$

When the phases are narrow distributed around a constant value, the S.I. goes to unity, otherwise for broad distributed phases it goes to zero. In Figure 5.11 it can be seen that the S.I. behaves similar to the amplitude at the Hopf values and shows a pronounced synchronization maximum over both, noise intensity and coupling strength. Coming closer the second Hopf–bifurcation point by further increase of noise intensity the S.I. decrease to zero. For higher the S.I. shrinks after passing the maximum. This lessening originates in the same reason given for the amplitude decrease.

Finally, both measures are able to detect and quantify the appearance of oscillations in the global output and their specification close to the Hopf–bifurcation values. However, both seem to fail for higher coupling values. Although the system is deep into to the oscillatory regime, the amplitude and the S.I. go down to zero. One reason for that is already given and concerns the separating time scales and with them the change in the oscillation characteristics. Nevertheless, the fact that the measures decay to zero is a finite size effect and would not occur within an ensemble including infinitely many units.

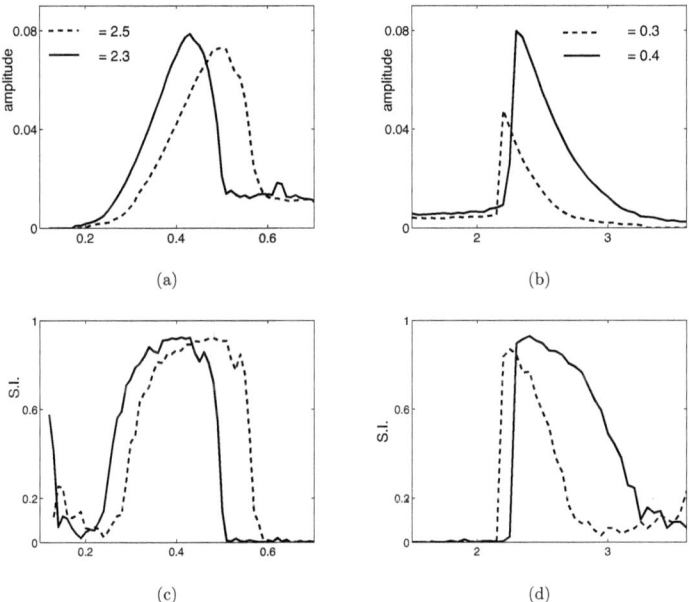

Figure 5.11: Oscillations amplitude (a-b) given in Eq. 5.30 and synchronization index (c-d) given in Eq. 5.32 with respect to noise intensity and coupling strength are plotted. Both measurements are shown for a range that involves the Hopf–bifurcations.

5.5 Conclusions

Coming from abstract models for excitable systems we construct a two–state model, containing one state for the stationary or polarized state and one state for the active or depolarized and refractory state. Both states are defined by characteristic waiting time distributions corresponding to excitable behavior. An ensemble of such two–state units is considered coupled by a feedback mechanism making the activation rate of single elements dependent on the fraction of active units within the ensemble.

In analytical studies we found for instantaneous feedback at the MF level monostability and bistability for the collective occupation probability distribution, which alternate through saddle–node or pitchfork bifurcations. Each unit changes between

5 Excitable two–state units coupled with delayed feedback

both states following its individual waiting time distribution. The latter is affected by a global excitatory coupling driving the ensemble to one of maybe multiple stationary stable states, but never to oscillatory behavior. It is important to note that although the ensemble reaches a stable state, the individual units still transit between excited and resting state, following their internal clock.

A more complex behavior arises, when global coupling feeds back to each individual activation time after a certain time delay. Then the ensemble passes through a Hopf–bifurcation to an oscillating regime, in the sense of almost synchronous activation of the individual units. A minimal time delay that creates oscillations in the global output is approximated and higher delay branches that appear recurrently are considered. For appropriate time delays a Hopf–bifurcation occurs also for inhibitory coupling, however for different values of noise intensity.

Finally, numerical simulations are performed to affirm analytical results and estimations. Two measures that capture oscillations best directly at the Hopf–bifurcation are introduced to quantify the periodic solutions and their synchronization level.

6 Synchronization and transport in an oscillating periodic potential

6.1 Introduction

In the previous chapter we have studied synchronization in an ensemble of excitable elements evoked by global delayed coupling. For this specific model we have shown, that the specific interplay of associated time scales decides whether collective behavior occur or not.

However, synchronization is a general and fundamental mechanism and can also emerge from the interplay of intrinsic dynamical time scales and external periodic driving. It has been investigated for numerous chaotic and stochastic dynamical systems. The ability to control those systems by mutual coupling or by external forces was demonstrated for various systems with a considerable number of possible applications [111, 3]. The great universality of the concept of synchronization is best seen in case of stochastic resonance [38, 85]. Here, in bistable and excitable systems even small periodic forces in combination with an optimally tuned noise are able to exhibit dynamical states which follow the oscillating periodic force.

For systems whose intrinsic dynamical time scales can be excited resonantly by driving noise, the effect of coherence resonance is established [85, 110]. Irregular spike trains produced by an excitable system can reach a state of high regularity and coherence at an optimal noise intensity. An early considered system besides neuron models, where coherence resonance was shown, has the form

$$\dot{\phi} = a - \cos\phi + \xi(t), \qquad (6.1)$$

where ϕ describes a phase variable and ξ may be any kind of noise [150]. For the same dynamics Stratonovich [137] described phase–locking between an external periodic driving and the oscillators orbit under the influence of noise (see section 6.5). That kind of dynamics further constitutes the nucleus of the famous Kuramoto equation [76]) on which the theory of synchronization for coupled phase oscillators was constructed.

Another aspect, we want to consider in this chapter concerns transport mechanisms within periodic fields. Interpreting ϕ as a spatial coordinate Eq. 6.1 also serves the modeling of such transport processes in which synchronization plays an important role. If the characteristic length scale is small enough omnipresent thermal

6 Synchronization and transport in an oscillating periodic potential

fluctuations can control the dynamics. The Brownian motion of particles in such periodic fields is one of the basic problems in applied stochastic dynamics. Escape processes over barriers in combination with diffusive motion within the potential valleys control the mean velocity as well as the dispersion of the transport [86, 121, 84, 12]. We will therefore make use of established quantities as effective diffusion and Péclet number, that characterizes the transport process.

One important representative of such systems with periodic fields are systems where particles move in periodic potentials and time dependent forces or modulations of the potential enhance a directed flow [120, 3, 50]. A recommendable compendium for these so called ratchets is given in [120]. Most of the research focused on mean drift within such systems and on the conditions under which the directed transport is maximized. Also the diffusion coefficient in these temporally changing potentials have been calculated [35] and it was proposed to take the diffusion coefficient to evaluate the precision of stochastic directed transport [89].

In this chapter we study overdamped Brownian transport in a temporally oscillating and spatially periodic potential. In [114] synchronized mobile states in stochastic ratchets have been reported where the force acts additively as moving periodic potential. Here we consider synchronization of Brownian particles in periodic potentials where the amplitude is modulated periodically. The interplay of the different time scales in the system, given by the period of the oscillating potential, the relaxation time of the deterministic dynamics and the diffusion time gives rise to non–trivial dynamical phenomena, such as an oscillation driven enhancement of the effective diffusion. Otherwise if additionally external forces are applied, the particle's motion becomes quasi–deterministic following the oscillations in direction of the applied force jumping several periods of the potential with minimal diffusion. The temporal modulations discussed here could be realized in different experimental systems such as free–flow dielectrophoresis [2], colloidal particles in optical fields [30, 140, 11], Josephson-junctions [66, 134] or paramagnetic colloids in magnetic fields [142, 5].

Note, that within the present chapter the notation of noise intensity and diffusion seems to be confused, because we denote both with D. The reason is simply, that for a spread of single trajectories in space caused by noise (Langevin equation level), the corresponding spatial probability distribution broadens due to a diffusion coefficient equals to the noise intensity (Fokker–Planck equation level).

6.2 Theta neuron

As a motivation and in order to demonstrate, that systems with an underlying periodic potential and excitable systems are structurally related, we will briefly discuss the Ermentrout–Kopell model [27]. It is an abstract one–component model similar to Eq. 6.1, that exhibits excitable and oscillatory behavior including one parameter being the bifurcation parameter for a saddle–node bifurcation that divides both dy-

6.2 Theta neuron

namical regimes. It may remind on the Ornstein–Uhlenbeck model, mentioned in chapter 2.2, in which the reccurence to the stationary state after deterministic or stochastic escape over an absorbing barrier is externally defined by a reset condition. The Ermentrout–Kopell model without such a reset condition describes the motion in an extended biased periodic potential.

The dynamical equation of the Ermentrout–Kopell model is given as

$$\dot{\Theta} = 1 - \cos\Theta + a(1 + \cos\Theta) = 1 + a - (1 - a)\cos\Theta, \tag{6.2}$$

with the associated potential

$$U(\Theta) = -(a + 1)\Theta + (1 - a)\sin\Theta. \tag{6.3}$$

For a variable Θ whose domain of definition contains every real number, Eq. 6.2 describes the motion of an overdamped particle in a periodic potential given by Eq. 6.3 and shown in the main figure of Fig. 6.1. That resembles typical transport processes in ratchets. For a restricted domain in which the running variable is defined on a circle with $-\ \leq \Theta \leq\ $, so that for the potential yields $U(\ ) = U(-\ )$ we find excitable or oscillatory behavior. The corresponding potential section is shown in the inset of Fig. 6.1. Once a particle has reached the right handed boundary at $\Theta =\ $ it is set to $\Theta = -\ $ per definition.

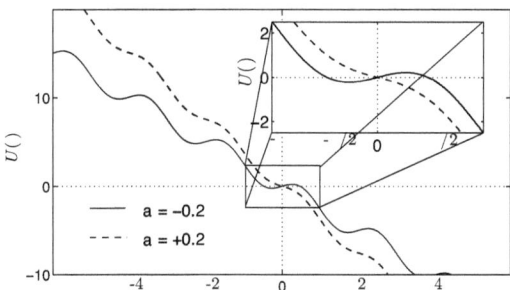

Figure 6.1: Two realizations of the potential from Eq. 6.3 of the Ermentrout–Kopell model. Over multiple periods it corresponds to a typical ratchet potential either including barriers or leading to sliding motion without obstruction controlled by a. The zoomed region shows a restricted area over $2\ $ which resembles excitable dynamics when periodic boundary conditions are taken.

6 Synchronization and transport in an oscillating periodic potential

The Ermentrout–Kopell model has two fixed points at

$$\Theta_\pm^0 = \pm \arccos\left(\frac{1+a}{1-a}\right) \in \mathbb{R} \text{ for } a \leq 0, \quad (6.4)$$

within the $[-\pi, \pi]$ range. The smaller value Θ_- is the rest state and is stable whereas Θ_+ corresponds to the threshold of the system and is unstable. For any $\Theta < \Theta_+$ the system decays back to the rest state. But once the threshold is passed the system runs toward the barrier and get reseted to $\Theta = -\pi$ corresponding to a large excursion in phase space and a spike in time course. Both fixed points merge for $a = 0$ where they annihilate due to a saddle–node bifurcation. Thus for $a > 0$ without having real fixed points the system has no attractor where the motion comes to rest. Due to the recurrent reset at the boundary an oscillatory dynamics is generated. The cases of a being smaller and greater than zero are exemplified in Fig. 6.1 as solid and dashed lines. As it can be seen the potential possesses barriers for $a < 0$ while for positive a no barrier obstruct the motion. The Ermentrout–Kopell model as an abstract neuron model is used for example in [101].

6.3 Dynamical equation

We consider an overdamped Brownian particle with friction Γ in a heat bath Q in a spatially periodic and biased potential $U(x,t)$ with oscillating amplitudes. The dynamics are described by the following Langevin equation

$$\dot{x} = -\frac{\nabla U(x,t)}{\Gamma} + \sqrt{\frac{2Q}{\Gamma}}\,\xi(t). \quad (6.5)$$

As in previous chapters the second term represents white Gaussian random force with zero mean and the intensity $2Q/\Gamma$ due to thermal fluctuations.

As the potential we choose $U(x,t) = A(x)B(t) - fx$, with f being a constant force, whereas $A(x) = A(x+\lambda)$ and $B(t) = B(t+T)$ are periodic functions in space and time with wavelength λ and period T. A reasonable and natural choice is the combination of two simple periodic functions

$$A(x) = -\frac{A_0}{2}\sin(qx), \quad B(t) = \sin(\Omega t). \quad (6.6)$$

Here A_0 is the maximal potential difference between two extrema for $f = 0$, $\Omega = 2\pi/T$ is the oscillation frequency and $q = 2\pi/\lambda$ is the spatial wave number. In contrast to typical ratchet models here the bias is kept constant in time whereas the amplitudes alternates their appearance as barriers and valleys. For the given potential Eq. 6.5 can be transformed into dimensionless variables $y = qx$, $\tilde{t} = \frac{A_0 q^2}{2}t$

6.3 Dynamical equation

and we obtain

$$\dot{y} = \cos(y)\sin(\Omega \tilde{t}) + F + \sqrt{2D}\,\xi(\tilde{t}), \qquad (6.7)$$

with dimensionless frequency $\Omega = 2\Gamma/(A_0 q^2)$, force $F = 2f/(A_0 q)$ and noise intensity $D = 2Q/A_0$. In the following the dimensionless time will be denoted as 't' for simplicity. The similarities between the dynamics from Eq. 6.7 and the Ermentrout–Kopell model appear to be obvious. Replacing the force with $F = a + 1$ and fixating the oscillating amplitude: $\sin(\Omega \tilde{t}) = a - 1$ our model transforms directly into the Ermentrout–Kopell model. However, regarding the explicit time dependency as an extra equation by setting $\dot{\phi} = \Omega$ it becomes clear, that the model defined by Eq. 6.7 has an extra dimension in phase space and furthermore, an extra degree of freedom due to an additional parameter.

Firstly, we are interested in transport properties which we investigate by considering the asymptotic drift velocity and for the mean spread of stochastic trajectories we examine the mentioned effective diffusion coefficient

$$v_{\text{drift}} = \lim_{t \to \infty} \frac{\langle y(t) \rangle - \langle y(0) \rangle}{t}$$
$$D_e = \lim_{t \to \infty} \frac{1}{2}\frac{d}{dt}\langle (y - \langle y(t) \rangle)^2 \rangle, \qquad (6.8)$$

where $\langle \rangle$ denotes the ensemble average. The quality of the directed transport, that occurs for $F \neq 0$, can be measured by the so called Péclet number (also sometimes referred to as Brenner number), defined as

$$\text{Pe} = \frac{v_{\text{drift}} L}{D_e}, \qquad (6.9)$$

where L is the characteristic length scale of the system given by $L = \lambda = 2$ for our system. For $\text{Pe} < 1$ diffusion dominates the dynamics and the directed transport plays a minor role in comparison with the non–directed spread of trajectories. For $\text{Pe} > 1$ the transport is dominated by the drift while the diffusion broadens the bunch of directed trajectories. The limit of $\text{Pe} \to \infty$ corresponds to a deterministic transport with vanishing effective diffusion on the characteristic length scale L.

Sample trajectories are shown in Fig. 6.3 for the case of $F = 0$ (a) and the biased case where $F > 0$ (b). They jump in steps of λ in space and in steps of $2\pi/\Omega \approx 31$ in time corresponding to the periodicity of both oscillating functions. Even with one trajectory for each case it is obvious, that for $F = 0$ no spatial direction is preferred while for $F = 0.1$ the trajectory moves forward only interrupted by a few back–jumps caused by noise. Note, that without noise, the trajectory moves perfectly directed in the direction the force is pointing because no permanent barriers exist in our model.

6 Synchronization and transport in an oscillating periodic potential

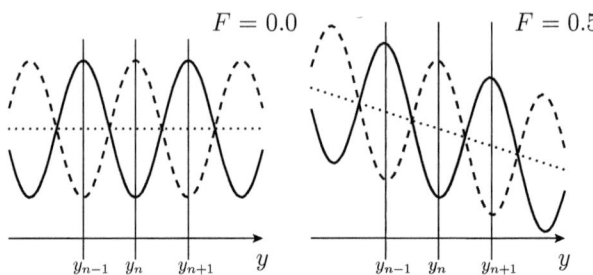

Figure 6.2: Illustration of the oscillating potential for $F = 0$ and $F = 0.5$ for $t = (n + 1)T$ (solid line), $t = (2n + 1)T/4$ (dotted line) and $t = (2n + 1)T/2$ (dashed line) with $n = 0, 1, 2, \ldots$.

6.4 Effective diffusion in case of $F = 0$

In case $F = 0$ the spatial and temporal symmetry in the potential prevents any directed flux within the system ($v_{\text{drift}} = 0$). In the following discussion we turn into the picture of propagating probability densities instead regarding single stochastic trajectories. We consider a sharp peaked Gaussian distribution initiated at the position y_0. The spread over the potential landscape is characterized by D_e, that may depend on the remaining system parameters D and Ω, in general.

For large noise intensities the influence of the potential is insignificant and the dynamics of the probability distribution is purely diffusive with the trivial limit $D_e \to D$. If D is sufficiently low temporarily occurring potential barriers prevent a too fast diffusion and the acting forces described through the deterministic part of the dynamics plays an essential role. Therein two different time scales controls the dynamics. On the one hand the external driving period $T = 2\pi/\Omega$ and secondly the intrinsic relaxation time t_r from an unstable potential maximum to a stable minimum. For oscillation frequencies much faster than this relaxation time $\Omega \gg 1/t_r$ another limiting case exists. Then the potential is self–averaged to be effectively flat and the probability distribution spreads with $D_e \to D$, likewise.

The potential, sketched in Fig. 6.2, turns its extrema within a half oscillation period $T/2 = \pi/\Omega$. A particle that follows the dynamics from Eq. 6.7 starting close to a maximum of the potential is able to approach the next nearest minimum, before it becomes transformed to a metastable maximum, only if the oscillation frequency is small compared to the inverse relaxation time $\Omega < \pi/t_r$. In that case the particle performs effectively discrete jumps (see Fig. 6.3 a) between the extreme states of the potential y_n, as labeled in Fig. 6.2. This oscillation induced random walk can be quantified by assuming the jump probability from a maximum to the left and

6.4 Effective diffusion in case of F = 0

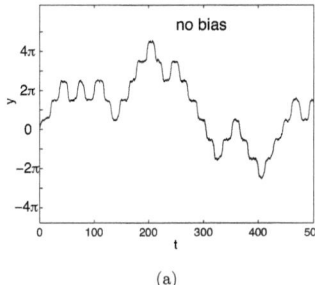 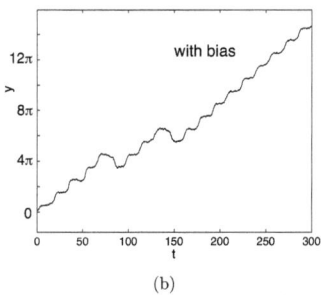

(a) (b)

Figure 6.3: Typical trajectories from Eq. 6.7. (a): $F = 0$ and (b): $F = 0.1$, remaining parameter values: $\Omega = 0.2$ and $D = 0.01$.

right as $p = 0.5$. The jumping events take place within discrete time intervals $T/2$ corresponding to the switching of the fixed pint's stability of the dynamics. The jump length is given by $l = /2 = $.

The probability to find the particle at the position $y_n = n$ after N jumps is given by the binomial distribution, which converges towards a Gaussian distribution in the long time limit

$$P(y,t) = \frac{1}{\sqrt{2D_e t}} \exp\left(-\frac{y^2}{2D_e t}\right). \tag{6.10}$$

The effective diffusion in the limit of many steps and small l with respect to the mean square displacement ($N \to \infty$, $l \to 0$) obey the relation $2D_e t = Nl^2$ with $N = 2t/T = t\Omega/$. By replacing $l = /2 = $ we obtain a linear relation between the frequency and the effective diffusion coefficient:

$$D_e(\Omega) = \frac{{}^2\Omega}{} = \frac{}{2}\Omega. \tag{6.11}$$

Thus, the faster the potential oscillates the faster the probability spreads over the potential. However, as discussed above for large frequencies the effective diffusion will reach asymptotically the noise intensity D, hence, the linear relation in Eq. 6.11 can not hold for the whole frequency range.

Obviously, the situations changes for a certain frequency value, at which the simple jump assumption fails. For intermediate frequencies the Gaussian distribution around a stable fixed point (potential minimum) y_n splits in two parts as soon as the fixed point becomes unstable (transforms into a potential maximum). Before the

6 Synchronization and transport in an oscillating periodic potential

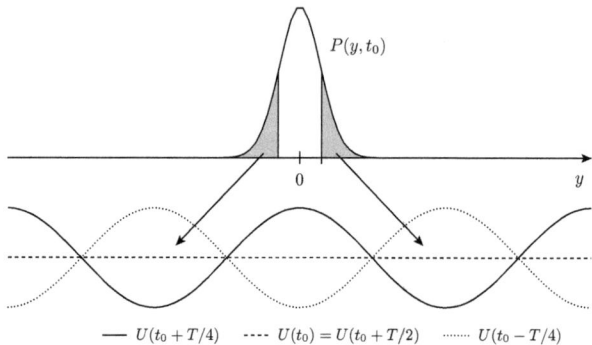

Figure 6.4: Schematic visualization of the ansatz for derivation of Eq. 6.15. We assume that only the gray shaded parts of the initial Gaussian distribution are able to reach the next temporal minimum within $T/2$.

split distribution relaxes completely at the neighboring fixed points $y_{n\pm 1}$ the oscillations switches the stability of the fixed points again and a part of the distribution moves back into the former position y_n. That befalls to the majority of the particles at higher oscillation frequencies Ω beyond a critical frequency which results in an effective localization of the distribution or a decreasing diffusion, respectively.

We attempt to describe this mechanism by estimating the fraction of probability $u(\Omega, D)$ of a initial distribution at y_n to reach $y_{n\pm 1}$ with respect to the oscillation frequency Ω, schematically depicted in Fig. 6.4. The initial position from which a particle is able to arrive the neighboring fixed point during half a period of oscillation can be expressed by $y_{\text{ini}} = y_n + \xi$ ($\xi \geq 0$), where ξ is a small displacement from the fixed point. Effects of noise are assumed to play only the qualitative role of small perturbations close to fixed points and we neglect noise effects during the sliding motion. Without loss of generality we set $y_n = 0$ and $y_{n+1} = \ell/2 = \delta$. The demanded probability is given by the complementary error function

$$u(\Omega, D) = \text{erfc}\left(\frac{\delta}{\sqrt{2\sigma^2}}\right), \qquad (6.12)$$

with the width σ^2 of the particle distribution at y_n at time t_0. Both the width and the displacement δ are assumed to become stationary after some initial transient time. In general σ^2 and δ are functions of Ω and D. By numerical simulations of the Eq. 6.7 it can be confirmed, that the width of the distribution is governed by the noise intensity and a simple ansatz $\sigma^2(D) = bD$ is reasonable. To obtain $\delta(\Omega)$

6.4 Effective diffusion in case of F = 0

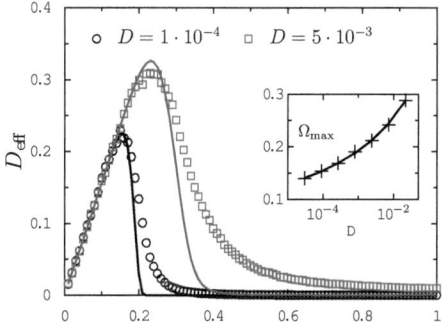

Figure 6.5: Effective diffusion vs. Ω obtained from simulations for $D = 1.0 \cdot 10^{-4}$ (circles) and for $D = 5.0 \cdot 10^{-3}$ (squares) in comparison with results of Eq. 6.15 (solid line) for $e = 0.09$ and $b = 150$. The inset shows the numerical results for the position $\Omega_{\max}$ of the maximum of D_e in dependence on D (crosses) and the prediction from Eq. 6.15 (solid line).

– noise effect are assumed to be negligible – we integrate the deterministic part of Eq. 6.7 over a half period

$$-\int^{y_{\text{end}}} \frac{dy}{\sin y} = \int_0^{T/2} \sin \Omega t \, dt. \qquad (6.13)$$

The left hand side would diverge for an upper limit at y_{end}. However, due to the noisy perturbations the particle is able to reach the next fixed point within a finite time. Therefore we set the final limit to $y_{\text{end}} = \ - \ $, where represents a small distance which the particle overcomes by fluctuations alone. By assuming to be Ω-independent we find an expression for the displacement from Eq. 6.13 as

$$= 2 \arctan\left[\exp\left(-\frac{2}{\Omega}\right) \cot\left(\frac{}{2}\right) \right]. \qquad (6.14)$$

Every particle with an initial displacement larger or equal to will reach the next fixed point or $y_{n+1} - $, respectively, within the half period of potential oscillations. Only the probability fraction u from Eq. 6.12 contributes to the effective diffusion, that reads therefore $D_e (\Omega, D) = \frac{}{} \Omega u(\Omega, D)$. As represents a small length and ^{2} is the variance of the initial Gaussian distribution, we assume their dependence on D as $^2 = bD$ and $= e\sqrt{D}$, where e, b remain undetermined. Thus we obtain

119

6 Synchronization and transport in an oscillating periodic potential

an expression for the effective diffusion, valid on a larger Ω range:

$$D_e\,(\Omega, D) = \frac{1}{2}\Omega\,\mathrm{erfc}\left(\frac{2}{\sqrt{bD}}\arctan\left[\exp\left(-\frac{2}{\Omega}\right)\cot\left(\frac{e\sqrt{D}}{2}\right)\right]\right). \qquad (6.15)$$

This result contains a more complex dependancy on Ω compared to the simple linear relation from Eq. 6.11 and, additionally, it depends on the noise intensity. However, it can be estimated, that for small frequencies the argument of the complementary error function collapses to zero and thus the linear limit still holds.

Considering the Ω dependencies further, we can see the result from Eq. 6.15 depicted in Fig. 6.5 as solid lines for $e = 0.09$ and $b = 150$ compared to numerical simulations (circles and squares). The expression in Eq. 6.15 clearly exhibits the supposed maximum in the effective diffusion for certain $\Omega = \Omega_{max}$ whose position depends on noise intensity D. After passing this maximum the effective diffusion decreases rapidly to very small values without reaching zero. So, the diffusive transport enhanced over two orders of magnitude by oscillations is efficient for a finite range of oscillation frequencies, where the time scale of intrinsic relaxation commensurate with the oscillation period of the potential.

The single choice for the values of the remaining free constants b and e is sufficient in order to cover the position of the oscillation frequency Ω_{max}, that maximizes the diffusion, over three orders of magnitude of D (see the inset in Fig. 6.5). Note, that the expression for D_e in Eq. 6.15 diverges for $\Omega \to \infty$. So, beyond frequency values considered here the presented approach fails.

Next, we will discuss the influence of moderate and high noise intensities. Therefore the quantity D_e/D is considered that quantifies the ratio of dynamical induced effects to pure Brownian motion. For $D_e/D = 1$ the presence of a potential does not affect the dynamics and trajectories correspond to pure Brownian particles, whereas for $D_e/D > 1$ the system is in the regime of enhanced diffusion induced by the potentials oscillation.

Low Ω and non–vanishing noise intensities yields $\to 0$ and therefore $D_e \approx \Omega/2$ = const. with respect to D. Thus over a large noise range the ratio D_e/D decreases as D^{-1}. That correspond to numerical results until a minimum where $D_e/D < 1$. (see black circles in Fig. 6.6). The ratio saturates for $D \to 0$ at larger Ω beyond the optimal frequency Ω_{max}, where the diffusion can not be enhanced over a certain limit (see gray pluses in Fig. 6.6). However, for the chosen Ω the effective diffusion remains about ten times larger than D even for very small noise intensities. A fit of Eq. 6.15 is in agreement with the numerical simulations only for not too low and not too large noise intensities. So smaller Ω increase the range of noise, in which Eq. 6.15 is a valid.

After a minimum is passed, where D_e/D sinks even below one, the ratio approaches asymptotically unity, where the potential is negligible, compared to the noise intensity. The minimum can be explained as a signature of barrier diffusion.

6.5 Synchronization in a biased potential

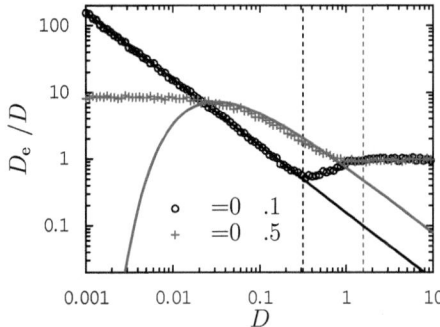

Figure 6.6: Simulation results of D_e/D vs D for $\Omega = 0.1$ (black symbols) and $\Omega = 0.5$ (gray symbols) compared with theoretical predictions for $\Omega = 0.1$: $D_e/D = \Omega/(2D)$ (black line) and for $\Omega = 0.5$ a fit of Eq. 6.15 (gray line). The dashed lines indicate $D_{\min}$ (Eq. 6.16).

With increasing noise the characteristic diffusion length per half-period of the oscillations approaches the characteristic length of the system $/2$. This leads to significant obstruction of the diffusive motion. An estimate for the location of the minimum $D_{\min}$ can be obtained from calculating the noise intensity D at which the mean square displacement of a free particle within the time $t = T/4$ equals $^2/4$:

$$D_{\min} = \frac{^2\Omega}{4} = \Omega,\qquad(6.16)$$

shown as dashed lines in Fig. 6.6.

Every numerical result presented in the current and in the next section are produced by simulations of the Eq. 6.7. We used a stochastic Heun algorithm to obtain single trajectories. We averaged over 30000 oscillation periods in time and over initial conditions $y_{\mathrm{ini}} \in (0,2\ )$ for computing the mean velocity and the effective diffusion. Additionally, an ensemble of 5-12 1024 trajectories are subsumed to get an appropriate statistics. The code was implemented on graphic cards using CUDA [61], which allows a high level of parallelization and thus an accelerated computation.

6.5 Synchronization in a biased potential

The addition of a constant force leads to a temporally constant tilt of the oscillating periodic potential. Two cases has to be distinguish. For $F > F_{\mathrm{crit}} = 1$ the drift motion of a particle is not hindered by the emergence of potential barriers at any

6 Synchronization and transport in an oscillating periodic potential

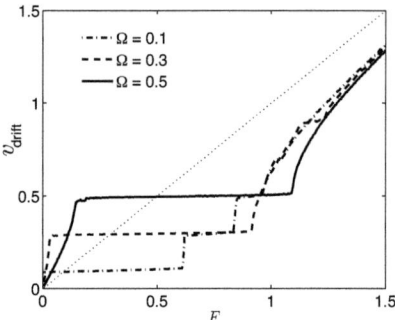

Figure 6.7: Deterministic drift velocity v_{drift} averaged over initial conditions as a function of the tilt shows different plateaus at $v_{\text{drift}} = \Omega, 3\Omega, 5\Omega, \ldots$. The dotted line stands for the potential–free case where $v_{\text{drift}} = F$.

time and contrariwise, there are potential barriers within time windows due to the oscillation period of the potential for $F < F_{\text{crit}}$. Note, that for both cases there is a finite drift in the direction of the bias F in contrast to static washboard potentials.

The drift velocity v_{drift} as a function of the bias shows different plateaus at $v_{\text{drift}} = \Omega, 3\Omega, 5\Omega, \ldots$, corresponding to a covered distance of 1 , 3 and 5 per oscillation period. This is shown in Fig. 6.7 for purely deterministic motion ($D = 0$) relative to the potential–free case where $v_{\text{drift}} = F$ (dotted line). For small bias until the the first plateau on which the motion is locked with $y = \Omega T = $, the drift velocity is even higher as in the potential–free case. In the plateau region a particle drifting down the potential is locked in a dynamical state. The oscillation frequency of the potential is synchronized with the frequency of / 2–displacements of the particle induced by the bias of the potential.

This behavior resembles characteristics of a driven oscillator, where such plateaus correspond to entrainment regimes of an oscillator to an external driving. The spatial coordinate of a particle jumping from one potential minimum to the next can be considered as the phase of a driven oscillator while the bias in our case corresponds to the external driving. As shown in [3] externally driven stochastic oscillators show a strong inhibition of the effective phase diffusion in the synchronized state, which agrees to the observed decrease of spatial diffusion in our system.

Based on these similarities, that were already mentioned introductorily for this chapter, we attempt to describe the dynamics of our system in vicinity of the synchronization regime by a corresponding phase oscillator. We change into the co–moving frame $z(t) = y(t) - \Omega t$ and average the deterministic dynamics over one

6.5 Synchronization in a biased potential

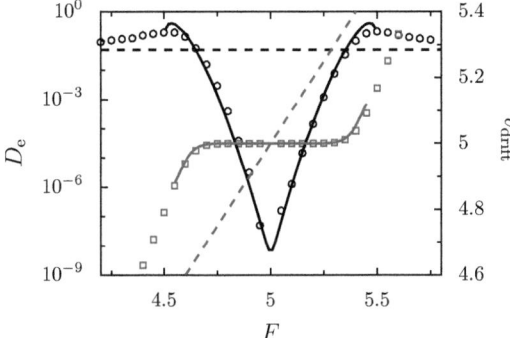

Figure 6.8: Analytical results for the drift velocity (gray solid line) and the effective diffusion (black solid line) from Eq. 6.19 and Eq. 6.20 at 1:1 synchronization and fast oscillations ($\Omega = 5.0$, $D = 0.05$) compared with simulation of Eq. 6.7 (squares: mean velocity, circles: effective diffusion). Dashed lines reference potential–free case.

oscillation period $T = 2/\Omega$. Due to the undetermined time–dependence of $z(t)$ we are not able to obtain a general solution of the averaged dynamics. However, we may write z as a Taylor series $z(t') = z(0) + z'(0)t' + \ldots$ with $t' = t/T$. Under the assumption that $z(t')$ changes slowly over one oscillation period we keep only the 0–th order from the series expansion and set $z(t') = z_T$. With this approximation we obtain the stochastic equation:

$$\dot{z}_T = \Delta - \frac{1}{2}\sin z_T + \sqrt{2D}\,\xi(t) \tag{6.17}$$

with $\Delta = |F - \Omega|$, which is in fact a specific form of Eq. 6.1.

Note, that due to the assumptions the equation holds only in the limit of fast oscillations with respect to the intrinsic relaxation time $\Omega \gg t_r^{-1}$. Thus for the locked state with 1:1 synchronization we have

$$\Omega \gg 1 \;\Rightarrow\; F = \Omega > F_{\text{crit}} = 1\,, \tag{6.18}$$

so we are in the super–critical force regime where no barriers obstruct the motion. In the co–moving frame even for that case the reduced Equation 6.17 describes a Brownian particle moving in a tilted stationary periodic potential relative to the bias caused drift from the original dynamics. An analytical solution for the mean drift velocity in such a system close to 1:1 synchronization was given by Stratonovich in

6 Synchronization and transport in an oscillating periodic potential

1967 [137]:

$$v_{\text{drift}} = w_0 \exp\left(-\frac{2}{D}w_0 - \frac{\Delta}{D}\arccos(2\Delta)\right) + \Omega, \quad (6.19)$$

with $w_0 = \frac{1}{2}\sqrt{1 - 4\Delta^2}$. The effective diffusion can be calculated in case of small noise compared to the barrier height in Eq. 6.17 ($D \ll 1/2$), from the Kramers rates $k^{\pm}$ to neighboring minima of the potential $U(z) = -\Delta z - \frac{1}{2}\cos(z)$ as $D_e = \frac{2}{2}\left(k^+ + k^-\right)$:

$$D_e = \frac{w_0}{2}\cosh\left(\frac{\Delta}{D}\right)\exp\left(-\frac{2}{D}w_0 - \frac{\Delta}{D}\arcsin(2\Delta)\right). \quad (6.20)$$

The effective diffusion breaks down at $\Delta = 0$ where the potential $U(z)$ is purely sinusoidal without bias. Symmetric barriers confines the stochastic particle close to a minimum and keep the corresponding probability density localized.

The dependency on F for both quantities is shown in Fig. 6.8 as solid lines. The potential–free case is presented as dashed lines and symbols stand for numerical simulations. Returning to original coordinates of the oscillating potential, the effective diffusion is minimized at $F = \Omega$. Around that resonant value the mean velocity passes a plateau at $v_{\text{drift}} = F = \Omega$, indicating the entrainment of a stochastic particle, that moves over one wavelength in the potential during one period.

Drift and diffusion together with the Péclet number is shown in Fig. 6.10 over a larger force range and for smaller oscillation frequencies. Again symbols represent numerical results using the expression from Eq. 6.9 to calculate the Péclet number. This quantity for the potential–free (dashed lines) case simply reads Pe = $2F/D$. For the two values of oscillation frequencies selected for Fig. 6.10, Ω = 0.1 and Ω = 0.05, the Stratonovich expressions from Eqs. 6.19, 6.20, do not hold due to the unfulfilled condition $\Omega \gg t_r^{-1}$. However, we present here an approximate approach to describe the dynamics for those smaller frequencies.

At low Ω (< 1) the first synchronization regimes are located at subcritical forces $F < 1$ (see Fig. 6.7), where the dynamics is dominated by oscillating barriers. A particle, that follows the potential has two modes of motion. First, close to a temporal minimum further directed drift down the bias is hindered by the head–on barrier mounting up for a finite time window in front of the particle. For that time interval the particle sticks and therefore we call it *stick* phase. When the barrier vanishes for half a period the particle moves down the biased potential and thus we call it *slip* phase.

We assume a –peaked probability distribution at the begin of the *slip* phase at the position y_0 corresponding to the fixed point where the particle is located at the beginning of the step. As soon as temporal barriers occur, the particle relaxes at $y_l = l/2$ with $l = (2n + 1)$ and $n \in \mathbb{N}$ and remains in the *stick* phase for $t = T/2$. After one oscillation parts of the distribution reach the next nearest minima and two

6.5 Synchronization in a biased potential

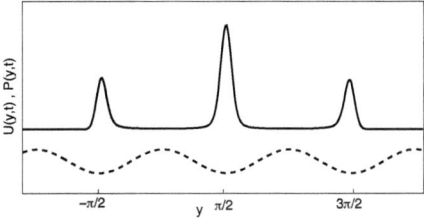

Figure 6.9: Snapshot of probability distribution (solid line) and corresponding potential (dashed line) after one slow oscillation period for an initiated δ-peaked distribution at $y_0 = \pi/2$.

'daughter peaks' are born as illustrated in Fig. 6.9 which are also narrow distributed around their minima.

The envelope of the generated distribution peaks is assumed to be Gaussian with a mean μ_e and the width S. The probability to find particles at unstable fixed points y_m with $m - 2n$ during the stick phase is insignificant small. The total probability to find the particle close to fixed point y_l at the end of the step reads

$$W_l = \frac{1}{2}\left[\text{erf}\left(\frac{y-\mu_e}{\sqrt{2S\Delta}}\right)\right]_{(l-1)/2}^{(l+1)/2} \quad (6.21)$$

Thus we can formally calculate the mean position and the variance after a single step

$$\langle y_e \rangle = \sum_{n=-\infty}^{\infty} y_{2n+1} w_{2n+1}, \quad (6.22)$$

$$\langle (y - \langle y_e \rangle)^2 \rangle = \sum_{n=-\infty}^{\infty} (y_{2n+1} - \langle y_e \rangle)^2 w_{2n+1}, \quad (6.23)$$

with $y_{2n+1} = (n + \frac{1}{2})\pi$. For that approximative approach it is sufficient to consider only a finite number of points $n \in [n_{min}, n_{max}]$ around μ_e and calculate the mean position and the variance by renormalizing the probabilities accordingly $w_{2n+1} = W_{2n+1}/\sum_{n=n_{m\,in}}^{n_{m\,ax}} W_{2n+1}$. With results obtained in Eqs. 6.22 and 6.23 we can calculate the mean drift and the effective diffusion as

$$v_{\text{drift}} = 2\langle y_e \rangle/T, \quad D_e = \langle (y - \langle y_e \rangle)^2 \rangle/T. \quad (6.24)$$

The mean of the particle probability distribution moves in the direction of the bias

6 Synchronization and transport in an oscillating periodic potential

with an effective force $G = G(F, \Omega)$ within a time interval $\Delta (F, \Omega)$ caused by the combined effect of the constant bias F and the oscillating potential. The time of vanishing barriers is $\Delta = 2\arcsin(F)/\Omega$ and for G we choose a linear dependence on F: $G(F) = F$. Averaging over the potential oscillations leads to = 1.

The diffusion coefficient S in Eq. 6.21 depends on the applied force and shows a large increase close to the critical force F_{crit} likewise reported in [121]. For simplicity we assume here a linear increase $S = F$, too. The simplified theory given in Eqs. 6.21, 6.24 is compared with numerical simulations of the system for $\Omega = 0.03, 0.05$ in Fig. 6.10. Symbols represent the numerical results while the black solid line represents the analytical calculation for a choice of fit parameter values: = 0.8 and = 0.04. For comparison the potential–free case is depicted as dashed lines again. The plateaus of the drift velocity, the corresponding strong decrease in the effective diffusion accompanied by multiple peaks of the Péclet number with decreasing magnitude are reproduced. For increasing Ω a shift of the peaks towards larger F is obtained which leads to a smaller number of peaks for $F < F_{\text{crit}}$. The most obvious difference of theoretical predictions and simulation data occurs for small F. By using = 0.8 we assume an overall reduced effective bias. However, as we have seen even for the deterministic case there is a small F range in which the addition of the oscillating potential accelerates the drift (see Fig. 6.7). In that range the fit parameter value should be chosen as > 1. We describe this through a step function around F_t as : $(F) = {}_2 + ({}_1 - {}_2)[\tanh((F_t - F)) - 1]/2$. This phenomenological modification of the model leads to the gray lines in Fig. 6.10 and shows agreement to the numerical simulations. Please note that F_t is the only changing parameter in calculations of the non–linear ansatz for different frequencies. This good agreement of the theoretical results with the numerical simulations shows that the theory accounts for the decisive mechanism responsible for the observed behavior: the combination of transport (*slip* phase) with repeated confinement (*stick* phase) at temporal minimum of the potential.

The giant Péclet numbers indicate a quasi–deterministic Brownian transport. For low noise strength ($D < 0.01$) and an optimal choice of F the system shows no dispersion even for extremely large times ($t > 10^5 T$, $\sim 10^4$ particles). Compared to the potential–free case effective diffusion shrinks and Péclet numbers rises over four order of magnitude.

6.6 Stationary probability density in the $y-$ plane

The dynamics given by Eq. 6.7 describes the course of single stochastic trajectories belonging to specific noise realizations. Instead, one may change to the equivalent description of probability density propagation, as we already did in the previous sections to investigate characteristic behavior. This consideration can help to clarify the appearance of alternating extrema pointed in Fig. 6.10 for D_e and Pe and give

6.6 Stationary probability density in the y − ϕ plane

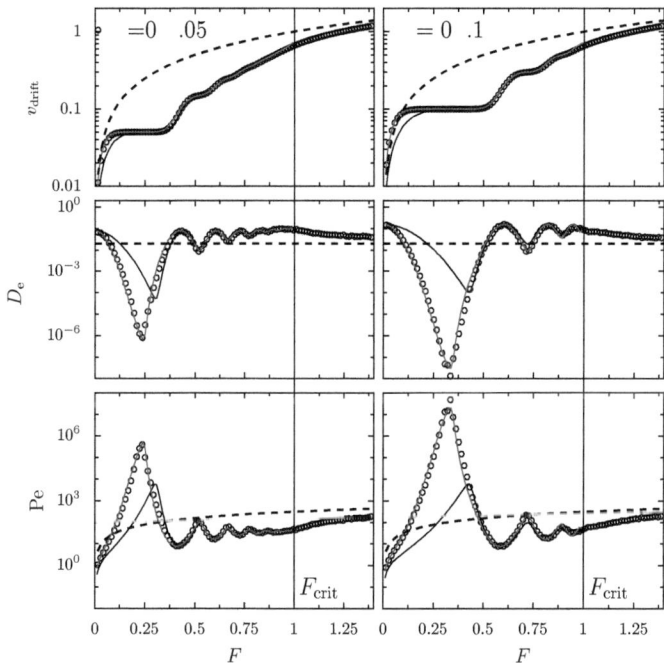

Figure 6.10: Comparison of simulations (symbols) with theoretical results at low Ω (solid lines): mean velocity v_{drift}, effective diffusion D_e and Péclet number Pe vs. bias F for $\Omega = 0.05$ (left column) and $\Omega = 0.1$ (right column) is shown. Black lines represent the linear ansatz $G = \gamma F$ with $\gamma = 0.8$, gray lines represent the non-linear ansatz with $G = \gamma(F)F$ with $\gamma_1 = 1.6$, $\gamma_2 = 0.8$, $\nu = 10$ for both frequencies. The only changing parameter is F_t: for $\Omega = 0.05$ $F_t = 0.28$, for $\Omega = 0.1$ $F_t = 0.4$. For both examples $\sigma = 0.08$ is chosen.

insights to the distribution of the probability to find a stochastic particle over the potential landscape.

By interpreting $\phi = \Omega t$ as a phase variable of the temporal oscillations, we can

127

6 Synchronization and transport in an oscillating periodic potential

rewrite Eq. 6.7 into the following autonomous two–component system

$$\dot\varphi = \Omega$$
$$\dot y = \cos(y)\sin(\varphi) + F + \sqrt{2D}\,\xi(t).$$ (6.25)

The ensemble of stochastic trajectories can be translated into the occupation density $p(y,\varphi;t)$ in the $y-\varphi$ plane as the two–dimensional representation of the probability shown in Fig. 6.4 or Fig. 6.9, respectively. The dynamics of $p(y,\varphi;t)$ due to the action of the potential and fluctuations in the space direction is given by the following Fokker–Planck equation:

$$\frac{\partial}{\partial t}p(y,\varphi;t) = \hat{L}\,p(y,\varphi;t) \quad \text{with}$$

$$\hat{L} = -\Omega\frac{\partial}{\partial \varphi} - \sin(\varphi)\frac{\partial}{\partial y}\cos(y) - F\frac{\partial}{\partial y} + D\frac{\partial^2}{\partial y^2}.$$ (6.26)

The nature of φ–dynamics causes periodic boundary conditions with 2π periodicity in φ. For the space coordinate we consider a ring with 4π periodicity, so that two minima in y exist. Then a stationary solution exists for Eq. 6.26 satisfying $\hat{L}p^0(y,\varphi) = 0$. The corresponding stationary probability currents follow from the equation of continuity and read:

$$J_\varphi^0(y,\varphi) = \Omega\,p^0(y,\varphi)$$
$$J_y^0(y,\varphi) = [F + \sin(\varphi)\cos(y)]\,p^0(y,\varphi) - D\frac{\partial}{\partial y}p^0(y,\varphi).$$ (6.27)

The current in φ–direction J_φ^0 is proportional to the amount of probability scaled by the frequency $\Omega > 0$. Since p^0 is positive or zero, J_φ^0 flows always in positive φ–direction. However, the probability current in y–direction is also determined by the negative gradient of p^0. This leads to a current which points away from probability maxima. Obviously, J_y^0 becomes larger and more directed in positive y–direction with increasing force F.

Eqs. 6.26 and 6.27 are solved numerically on a adjusted grid using the finite element method. For $\Omega = 0.1$ and $D = 0.001$ and four force values the resulting probability densities and currents are shown in Fig. 6.11. The amount of probability is coded as gray levels, where increasing blackness represents accumulation of probability. Additionally, black arrows are included displaying the direction and strength of probability currents and black crosses mark the stable fixed point locations of the system due to Eqs. 6.25.

In the symmetric case ($F = 0$) probability is accumulated where stable fixed points according to minima of the potential are located, discernable as red stripes in the figure. From these probability maxima the current splits symmetrically to

6.6 Stationary probability density in the $y-$ plane

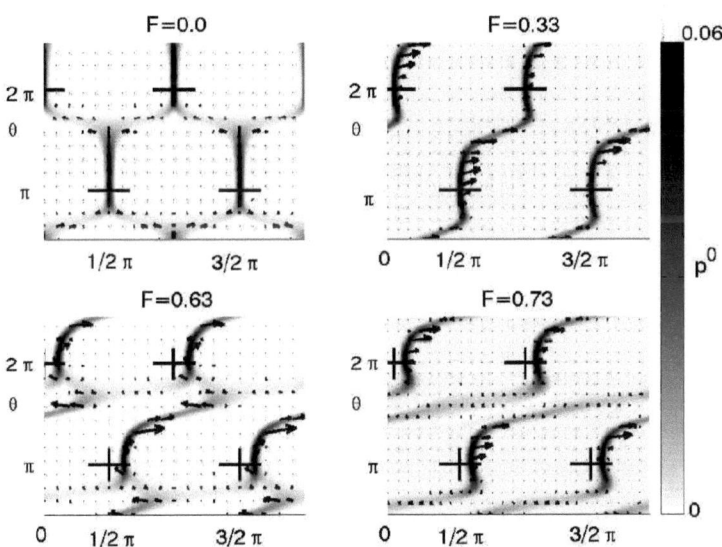

Figure 6.11: Stationary probability density $p^0(y,\ )$ coded as gray levels from numerical simulations of Eq. 6.26 is shown within the phase space range which is periodically continued for different forces F. Black arrows indicate the vector field of the probability currents J_y^0 and J^0. The black crosses locate the stable fixed points of the underlying dynamics. Parameter values are $\Omega = 0.1$ and $D = 0.001$. (Colors online)

6 Synchronization and transport in an oscillating periodic potential

the left and right. After half a period $\Delta =$ the probability gathers at the next minima of the potential.

The next three values of F used in Fig. 6.11 correspond to:

o the first maximum of the Péclet number at $F \approx 0.33$,

o the first minimum of Pe at $F \approx 0.63$,

o and the second maximum of Pe at $F \approx 0.73$

that are shown in Fig. 6.10. The probability distribution corresponding to 1:1 synchronization ($F \approx 0.33$) shows clearly the direct thin path from one fixed point to the neighboring each $\Delta =$. This path keeps localized and has a very small dispersion which is manifested in a vanishing effective diffusion coefficient and a giant Péclet number. For $F \approx 0.63$ the paths split and spread to the left and right hand sided fixed points, leading to a large effective diffusion. In this case v_{drift} is in between two plateaus. The distorted paths in the lower left figure clearly illustrates the asynchronous state. Finally, the fourth picture in Fig. 6.10 shows the case of $F \approx 0.73$ at the second maximum of Pe, corresponding to the second minimum of D_{e} . The probability path is extended, connecting every second minimum within a half period and thus overleaping one stable fixed point, according to 1:3 synchronization. The reason, why the second peak is less pronounced at this synchronization level is visualized here. Paths spreads stronger when they cover the doubled distance while the potential is relatively flat, which leads to higher effective diffusion.

6.7 Conclusions

In this last chapter of this work we treat a dynamical system that is structurally similar to excitable systems. We have analyzed the non–directed and directed transport of Brownian particles moving in an a spatially periodic potential with oscillating amplitudes. In the case of vanishing bias the oscillating potential may enhance the effective diffusion for a specific frequency range. We derived an expression for D_{e} (Ω, D) to quantify the enhancement which gives us an optimal driving frequency which maximizes the diffusive transport.

The introduction of a finite bias leads to a finite drift velocity due to broken symmetry and additional characterizing quantities besides effective diffusion are used. Particularly by considering drift velocity and Péclet number we found synchronization regimes, with strongly suppress effective diffusion and quasi–deterministic transport at finite noise strengths. For fast oscillations analytical expression based on previous results by Stratonovich and Kramers are found for the effective diffusion and mean velocity to quantify the synchronized state. Also for low frequencies we gave explanations for the appearance of higher order synchronization. Especially the study of probability densities and probability currents offered an approximative

6.7 Conclusions

access to analyze drift velocity, effective diffusion and Péclet number over a large frequency range. The latter approach brought us to the investigation of probability density distribution on a two–dimensional phase space containing the spatial direction and the phase variable of the temporally oscillatory state. Paths of stationary probability density obtained by numerical simulations gave more specific insights in the dynamical appearance of synchronized or desynchronized states.

Our results suggest that in the unbiased case the introduction of an spatio–temporal periodic force mainly enhances the diffusive transport, whereas for a finite bias it may significantly suppress diffusion and improve the quality of directed Brownian transport.

7 Summary

In this thesis spatially extended dynamical systems are studied that are under the influence of noise. Generally the combination of nonlinear dynamics, noise and spatially distribution can lead to a high grade of complexity. We have studied in this work a type of models that have a degree of transparency which allowed us to obtain essential modes of their dynamical characteristics. Problems addressed in this work ranging from noise induced activation of a single excitable unit over pattern formation in a heterogeneous active medium and abstract excitable two-state units coupled to their own output up to transport properties of Brownian particles moving in a periodic oscillating potential. Even if these scenarios seem to come from different fields, their qualitative properties are basically describable by the same type of models.

We will briefly list the most important results here, to give an overview of the most important questions and results: The stochastic escape from a fixed point is studied in chapter 2 by applying the FHN model as a prototype of an excitable system. The scaling of the mean escape time for major parameters is shown for a single unit in Fig. 2.4 and for a unit which is coupled to inactive neighbors in Fig 2.9. The latter is considered to be the very first initialization of a nucleation event. We noticed a qualitative difference in the mean escape time characteristics for a medium in the excitable regime compared to the subexcitable case.

An extension of the two–component FHN model is analyzed in the subsequent chapter 2. An extra equation was added in order to model the rising potassium ion concentration observed in extracellular space as the surrounding of excitable cells. We referred to the set of three equations as FNK model. The coupling of an excitable element to the external variable led to a rich dynamics such as excitable, oscillatory and bistable behavior as well as their coexistence as shown in the bifurcation diagram Fig. 3.3. The influence of noise on this three component situation was analyzed by determining the power spectrum and ISI distribution (see Fig. 3.7). The main focus of this chapter laid on spatially extended patterns emerging in a heterogeneous medium that consists of active units embedded in an inactive environment. By variation of parameter values that are responsible for the exterior dynamics we found well known structures such as wave fronts, spirals (Fig. 3.10) and target patterns (Fig. 3.13) as well as unusual structures such as noise supported traveling clusters (Fig. 3.11) and inverted spiral waves (Fig. 3.15). A survey of these patterns classified by their parameter dependent appearance is shown in Fig. 3.17. Thus we were able to produce a rich zoo of different patterns which are partly known

7 Summary

from measurements in real biological systems. The diversity was explained by the underlying dynamics in combination with the presence of noise. Which type of patterns develop in an active medium depends also on boundary conditions.

This is discussed in detail for a bistable front interacting with the system boundaries in chapter 4. We concentrated our analysis on Dirichlet boundary conditions for which we had to distinguish two kinds of interaction, basically. Either the front reflects on the boundary or it get bound and stops moving. The types of possible boundary interaction are subsumed in Fig. 4.6. First we considered a discrete array consisting of bistable elements coupled in the activator for which we obtained expressions that yields the critical parameter values where the dynamical regime changes. This is summarized in a parameter plane in Fig 4.8. Then we extend our considerations to the continuous bistable wave front and analyzed first diffusion in the activator with an immobile inhibitor dynamics and second diffusion in both variables. For the first case we find approximated expressions for the wave shape (Eqs. 4.21) and the front velocity by series expansions for small velocities and for a small time scale separation parameter. For the second case we linearized the problem and found a relation between the front velocity and the distance of the front to the boundary given in Eq. 4.43 which can be also applied for the Neumann boundary problem. Trajectories obtained by numerical simulations, that follows approximately nullclines defined in the distance–velocity phase space are shown in Figs. 4.16– 4.18. Consequently, the origin of reflexion of fronts at boundaries has been explained as a mechanism of metastability of the stationary states.

An abstract excitable system modeled as units that possess two discrete states is studied in chapter 5. They are coupled by their common output which lowers the activation rate of single units. This kind of positive feedback mechanism is connected to the three–component FNK model where the external ion concentration enhances the activity of the embedded neuronal units, however we dealt with an all–to–all connection in this chapter. The coupling via the activation rate of the two–state units leads to stationary states of the ensemble given by Eq. 5.13. We obtained mono– and bistability for the stationary mean occupation probability to be in the upper active state or in the lower rest state as shown in Fig. 5.6. If the rate of single units reacts on the common output from a former point in time then the coupling is delayed. For this case the ensemble's activity can be oscillatory for a certain parameter range (see Fig. 5.7) and we find regimes of synchronization. Again, coupling of active units led to the emergence of coherent patterns, which were manifested as structures in time in this case.

The problem we addressed in the last chapter 6 has a high intrinsic periodicity in space and time and is also connected to synchronization phenomena. We analyzed the motion of a Brownian particle in a spatio–temporal oscillating potential. The model we used is comparable to active units in their oscillatory phase as considered in previous chapters. However, the main aspect here laid on transport properties which we quantified via the averaged velocity and the effective diffusion as typical observables. Without a symmetry–breaking bias we obtained an expression for the effective diffusion given in Eq. 6.15 which was maximized by a finite oscillation frequency (see Fig. 6.5). For a tilted potential for which one direction of propagation is preferred, velocity plateaus over the tilting force appear where the motion of the particles is locked due to synchronization between the oscillation period and the time of sliding into the potential valleys. Interpreting this system as a driven oscillator we used expressions for the mean velocity: Eq. 6.19 and effective diffusion: Eq. 6.20 to describe the synchronized states for supercritical tilts. An approximated quantitative description is also given for smaller tilts that we compared to numerical results as shown Fig. 6.10. The interplay of deterministic dynamics and noise that produced coherent patterns in the first chapters led here to a coherent transport through the system due to similar synchronization effects. The transport paths can be interpreted as structures in the corresponding phase space as shown Fig. 6.11.

To summarize; we studied the interplay of nonlinear dynamics and noise for spatially extended and coupled active systems. We found transitions from one dynamical regime to another and effects of synchronization as an underlying mechanism for coherent pattern formation in space and time.

Appendix A

1 Mean first passage time

The density distribution $p(x,t)$ for the occupation probability of a Brownian particle in a one–dimensional potential $U(x)$ with a damping constant set to one is given by the Langevin dynamics

$$\dot{x}(t) = -U'(x) + \sqrt{2}\,\xi(t)$$

is described by the corresponding Fokker-Planck equation

$$\dot{p}(x,t) = \frac{\partial}{\partial x}\left[U'(x)p(x,t)\right] + \frac{\partial^2}{\partial x^2}p(x,t) = -\frac{\partial}{\partial x}J,$$

with D as the diffusion coefficient and J denotes the probability current. It follows

$$J = -\left[\frac{U'(x)}{}p + p'\right],$$

where the bar means derivation with respect to x. Multiplying with the Arrhenius factor leads to

$$J = -e^{-\frac{U}{\sigma}}\left[\frac{U'(x)}{}e^{\frac{U}{\sigma}}p + e^{-\frac{U}{\sigma}}p'\right] = -e^{-\frac{U}{\sigma}}\frac{\partial}{\partial x}\left[e^{\frac{U}{\sigma}}p\right].$$

After integrating and writing the current J as a flow of probability per time $J = W/T$ we have

$$T\int_a^b dx\,\frac{\partial}{\partial x}\left[e^{\frac{U}{\sigma}}p(x,t)\right] = Te^{\frac{U(b)}{\sigma}}p(b,t) - Te^{\frac{U(a)}{\sigma}}p(a,t)$$

$$= -\frac{1}{}\int_a^b dx\,e^{\frac{U(x)}{\sigma}}W(x,t) = -\frac{1}{}\int_a^b dx\,e^{\frac{U(x)}{\sigma}}\int_{-\infty}^x dy\,p(y,t).$$

We set b as the absorbing boundary where the probability density is zero ($p(b,t) = 0$). For an a close to a potential's minimum which is deep and far from the absorbing boundary the probability density in the last integrand can be approximatly estimated relative to the probability around a as $p(y,t) = p(a,t)\exp(\Delta U/\,)$ with a potential difference $\Delta U = U(x) - U(a)$. Since the current J is constant in space, T depends only on the distance over which probability is transported, given by a

Appendix A

and b. So, we arrive at the expression for the mean time to reach the boundary b stochastically when started in a:

$$T = \frac{1}{\,} \int_a^b dx\, e^{U(x)/\,} \int_{-\infty}^x dy\, e^{-U(y)/\,} \;.$$

Appendix B

2 Stationary solutions in the globally coupled two–state system

The balance of probability flows serves to determine the occupation probabilities $P_i(t)$, $i = 1, 2$ of two separate states i. The generalized master equations that hold for these probabilities read:

$$\frac{d}{dt}P_i(t) = J_j(t) - J_i(t), \qquad (1)$$

where $J_i(t)$ denote the probability flow from state i to j at time t. They can be expressed by the convolution of the inflow J_j to the state i in the past and the time distribution to wait in this state:

$$J_i(t) = \int_0^\infty dt'\, J_j(t-t') w_i(t'). \qquad (2)$$

Laplace transformation of the convolution gives the product:

$$\hat{J}_i(u) = \hat{J}_j(u)\hat{w}_i(u) \qquad (3)$$

and therefore the balance equation (1) transforms to

$$u\hat{P}_i(u) = \hat{J}_i(u)\hat{w}_j(u) - \hat{J}_j(u)\hat{w}_i(u), \qquad (4)$$

The waiting time distributions are normalized and can be splitted in parts:

$$1 = \int_0^\infty w_i(t')\,dt' = \int_0^t w_i(t')\,dt' + \int_t^\infty w_i(t')\,dt'. \qquad (5)$$

The survival probability can be introduced as the counterpart to the waiting probability

$$z_i(\ ) = 1 - \int_0^t w_i(t')\,dt' = \int_t^\infty w_i(t')\,dt' \qquad (6)$$

and after Laplace transformation it reads

$$\hat{z}_i(u) = \frac{1}{u}\bigl(1 - \hat{w}_i(u)\bigr). \qquad (7)$$

Appendix B

Using Eqs. 3, 4 and 7 we write for the probability to be in state i:

$$\hat{P}_i \underset{(4)}{=} \frac{\hat{J}_i \hat{w}_j}{u} - \frac{\hat{J}_j \hat{w}_i}{u} \underset{(3)}{=} \frac{1}{u}(1 - \hat{w}_i)\hat{J}_j \underset{(7)}{=} \hat{z}_i \hat{J}_j. \tag{8}$$

Backward Laplace transformation yields

$$P_i(t) = \int_0^\infty dt'\, J_j(t - t')\, z_i(t'). \tag{9}$$

Since the transition $1 \to 2$ is a rate process, its probability flow is given by $J_1(t) = P_1(t)$. Then equation (9) expressed for state 2 reads:

$$P_2(t) = \int_0^\infty dt'\, (1 - P_2(t - t'))\, z_2(t'). \tag{10}$$

Let P_2^* be the stationary solution of Eq. 10. Then we have

$$P_2^* = (1 - P_2^*) \int_0^\infty dt'\, z_2(t'). \tag{11}$$

Partial integration over the survival probability yields

$$\int_0^\infty dt\, z_2(t) = \int_0^\infty dt \int_t^\infty w_2(t')\, dt' = \underbrace{t\left[\int_t^\infty w_2(t')\, dt'\right]\Big|_0^\infty}_{=0} - \int_0^\infty dt\, t \frac{d}{dt} \int_t^\infty w_2(t')\, dt'$$

$$= -\int_0^\infty dt\, t \frac{d}{dt}\left(1 - \int_0^t w_2(t')\, dt'\right) = \int_0^\infty dt\, t\, w_2(t) = t_2. \tag{12}$$

Then the stationary solution can be written as the fraction of residence time to be in state 2:

$$P_2^* = \frac{t_2}{1/\ + t_2} = \frac{t_2}{T}, \tag{13}$$

where T denotes the time of a complete cycle.

Bibliography

[1] K. Agladze, J. P. Keener, S. C. Müller, and A.Panfilov. Rotating spiral waves created by geometry. *Sciene*, 264:1746–1748, 1994.

[2] A. Ajdari and J. Prost. Free-flow electrophoresis with trapping by a transverse inhomogeneous field. *Proc. Natl. Acad. Sci. U.S.A.*, 88:4468, 1991.

[3] V. S. Anishchenko, V. Astakhov, A. Neiman, T. Vadivasova, and L. Schimansky-Geier. *Nonlinear Dynamics of Chaotic and Stochastic Systems*. Springer, Berlin, 2007.

[4] F. M. Atay and A. Hutt. Neural fields with distributed transmission speeds and long-range feedback delays. *SIAM J. Appl. Dyn. Syst.*, 5:670, 2006.

[5] A. Auge, A. Weddemann, F. Wittbracht, and A. Hutten. A hydrodynamic switch: Microfluidic separation system for magnetic beads. *Appl. Phys. Lett.*, 94:183507, 2009.

[6] M. Bär, N. Gottschalk, M. Eiswirth, and G. Ertl. Spiral waves in a surface reaction: Model calculations. *J. Chem. Phys.*, 100:1202, 1994.

[7] M. Bär and G. Or-Guil. Alternative scenarios of spiral breakup in a reaction-diffusion model with excitable and oscillatory dynamics. *Phys. Rev. Lett.*, 82:1160–1163, 1999.

[8] R. Báscones, J. García-Ojalvo, and J. M. Sancho. Pulse propagation sustained by noise in arrays of bistable electronic circuits. *Phys. Rev. E*, 65:061108, 2002.

[9] M. Bazhenov, I. Timofeev, M. Steriade, and T. J. Sejnowski. Potassium model for slow (2-3 Hz) neocortical paroxysmal oscillations in vivo. *J. Neurophysiol.*, 92:1116–1132, 2004.

[10] I. Berenstein, M. Dolnik, L. Yang, A. M. Zhabotinsky, and I. R. Epstein. Turing pattern formation in a two-layer system: Superposition and superlattice patterns. *Phys. Rev. E*, 70:046219, 2004.

[11] S. Bleil, P. Reimann, and C. Bechinger. Directing Brownian motion by oscillating barriers. *Phys. Rev. E*, 75:031117, 2007.

Bibliography

[12] M. Borromeo and F. Marchesoni. Resonant transport in pulsated devices: Mobility oscillations and diffusion peaks. *Phys. Rev. E*, 78:051125, 2008.

[13] H. Brandtstädter, M. Braune, I. Schebesch, and H. Engel. Experimental study of the dynamics of spiral pairs in light-sensitive Belousov-Zhabotinskii media using an open-gel reactor. *Chem. Phys. Lett.*, 323:145–154, 2001.

[14] R. Brown. Additional remarks on active molecules. *Philos. Mag. N. S.*, 6:161–166, 1829.

[15] J. Buck and E. Buck. Mechanism of rhythmic synchronous flashing of fireflies fireflies of southeast asia may use anticipatory time-measuring in synchronizing their flashing. *Science*, 159-3821:1319–1327, 1968.

[16] S. Coombes and Y. Timofeeva. Sparks and waves in a stochastic fire-diffuse-fire model of Ca^{2+} release. *Phys. Rev. E*, 68:021915, 2003.

[17] J. R. Jr. Cressman, G. Ullah, J. Ziburkus, S. J. Schiff, and E. Barreto. The influence of sodium and potassium dynamics on excitability, seizures, and the stability of persistent states: I. single neuron dynamics. *J. Comput. Neurosci.*, 26:159–170, 2009.

[18] M. C. Cross and P. C. Hohenberg. Pattern formation outside of equilibrium. *Rev. Mod. Phys.*, 65:851–1112, 1993.

[19] M. A. Dahlem, G. Hiller, A. Panchuk, and E. Schöll. Dynamics of delay-coupled excitable neural systems. *Int. J. Bifur. Chaos*, 19:745, 2009.

[20] Y. A. Dahlem, M. A. Dahlem, K. Braun T. Mair, and S.C. Müller. Extracellular K^+ accumulation in the central nervous system. *Exp. Brain Res.*, 152:221, 2003.

[21] P. Dayan and L. F. Abbott. *Theoretical Neuroscience - Computational and Mathematical Modeling of Neural Systems*. The MIT Press Cambridge, Massachusetts, 2001.

[22] J. W. Deitmer, C. R. Rose, T. Munsch, J. Schmidt, W. Nett, N.-P. Schneider, and C. Lohr. Leech giant glial cell: Functional role in a simple nervous system. *Glia*, 28:175, 1999.

[23] M. Dhamala, V. K. Jirsa, and M. Ding. Enhancement of neural synchrony by time delay. *Phys. Rev. Lett.*, 92:074104, 2004.

[24] I. Dietzel, U. Heinemann, and H. D. Lux. Relations between slow extracellular potential changes, glial potassium buffering, and electrolyte and cellular volume changes during neuronal hyperactivity in cat brain. *Glia*, 2 (1):25–44, 1989.

Bibliography

[25] I. R. Efimov, V. Sidorov, Y. Cheng, and B. Wollenzier. Evidence of three-dimensional scroll waves with ribbon-shaped filament as a mechanism of ventricular tachycardia in the isolated rabbit heart. *Journal of Cardiovascular Electrophysiology*, 10:1452–1462, 1999.

[26] I. R. Epstein and V. K. Vanag. Complex patterns in reactive microemulsions: Self-organized nanostructures? *Chaos*, 15:047510, 2005.

[27] G. B. Ermentrout and N. Kopell. Parabolic bursting in an excitable system coupled with a slow oscillation. *SIAM J. Appl. Math.*, 46:223, 1986.

[28] M. Falcke. Deterministic and stochastic models of intracellular Ca^{2+} waves. *New J. Phys.*, 5:96, 2003.

[29] M. Falcke, L. Tsimring, and H. Levine. Stochastic spreading of intracellular Ca^{2+} release. *Phys. Rev. E*, 62:2636–2643, 2000.

[30] L. P. Faucheux, L. S. Bourdieu, P. D. Kaplan, and A. J. Libchaber. Optical thermal ratchet. *Phys. Rev. Lett.*, 74:1504, 1995.

[31] S. Fauve and F. Heslot. Stochastic resonance in a bistable system. *Phys. Lett. A*, 97-1/2:5–7, 1983.

[32] A. P. Fertziger and J. B. Jr. Ranck. Potassium accumulation in interstitial space during epileptiform seizures. *Exp. Neurol.*, 26(3):571–581, 1970.

[33] R. A. Fisher. The wave of advance of advantageous genes. *Ann. Eugenics*, 7:353–369, 1937.

[34] R. FitzHugh. Impulses and physiological states in theoretical models of nerve membrane. *Biophysical Journal*, 1(6):445–465, 1961.

[35] J. A. Freund and L. Schimansky-Geier. Diffusion in discrete ratchets. *Phys. Rev. E*, 60:1304, 1999.

[36] F. Fröhlich, M. Bazhenov, V. Iragui-Madoz, and T. J. Sejnowski. Potassium dynamics in the epileptic cortex: New insights on an old topic. *Neuroscientist*, 14:422, 2008.

[37] F. Fröhlich, M. Bazhenov, I. Timofeev, M. Steriade, and T. J. Sejnowski. Slow state transitions of sustained neural oscillations by activity dependent modulation of intrinsic excitability. *J. Neurosci.*, 26(23):6153–6162, 2006.

[38] L. Gammaitoni, P. Hänggi, P. Jung, and F. Marchesoni. Stochastic resonance. *Rev. Mod. Phys.*, 70:223, 1998.

Bibliography

[39] A. R. Gardner-Medwin. Analysis of potassium dynamics in mammalian brain tissue. *J. Physiol.*, 335:393–426, 1983.

[40] W. Gerstner and W. M. Kistler. *Spiking Neuron Models: Single Neurons, Populations, Plasticity.* Cambridge University Press, 2002.

[41] J. C. Goldstein and M. O. Scully. Nonequilibrium properties of an Ising-model ferromagnet. *Phys. Rev. B*, 7:1084–1096, 1973.

[42] J. D. Green. The hippocampus. *Physiol. Rev.*, 44:561–608, 1964.

[43] G. T. Gurija and M. A. Livshits. Nonequilibrium spatio-temporal selforganization due to delayed negative feedback. *Z. Phys. B - Condensed Matter*, 47:71, 1982.

[44] G. Haas, M. Bär, I. G. Kevrekidis, P. B. Rasmussen, H. H. Rotermund, and G. Ertl. Observation of front bifurcations in controlled geometries: From one to two dimensions. *Phys. Rev. Lett.*, 75:3560–3563, 1995.

[45] A. Hagberg and E. Meron. Pattern formation in non-gradient reaction-diffusion systems: the effects of front bifurcations. *Nonlinearity*, 7:805, 1994.

[46] D. Haim, G. Li, Q. Ouyang, W. D. McCormick, H. L. Swinney, A. Hagberg, and E. Meron. Breathing spots in a reaction-diffusion system. *Phys. Rev. Lett.*, 77:190–193, 1996.

[47] H. Haken. *Brain Dynamics: Synchronization and Activity Patterns in Pulse-Coupled Neural Nets with Delays and Noise.* Springer, Berlin, 2006.

[48] S. K. Han, T. G. Yim, D. E. Postnov, and O. V. Sosnovtseva. Interacting coherence resonance oscillators. *Phys. Rev. Lett.*, 83:1771, 1999.

[49] E. Hanert, E. Schumacher, and E. Deleersnijder. Front dynamics in fractional-order epidemic models. *J. Theor. Biol.*, 279-1:9–16, 2011.

[50] P. Hänggi and F. Marchesoni. Artificial brownian motors: Controlling transport on the nanoscale. *Rev. Mod. Phys.*, 81:387, 2009.

[51] A. J. Hansen. The extracellular potassium concentration in brain cortex following ischemia in hypo- and hyperglycemic rats. *Acta Physiol. Scand.*, 102:324, 1978.

[52] H. Hempel, L. Schimansky-Geier, and J. García-Ojalvo. Noise-sustained pulsating patterns and global oscillations in subexcitable media. *Phys. Rev. Lett.*, 82:3713, 1999.

Bibliography

[53] H. Henry and H. Levine. Wave nucleation rate in excitable systems in the low noise limit. *Phys. Rev. E*, 68:031914, 2003.

[54] A. F. Hodgkin and A. L. Huxley. A quantitative description of membrane currents and its application to conduction and excitation in nerve. *Journal of Physiology*, 117:500–544, 1952.

[55] D. Huber and L. S. Tsimring. Dynamics of an ensemble of noisy bistable elements with global time delayed coupling. *Phys. Rev. Lett.*, 91:260601, 2003.

[56] D. Huber and L. S. Tsimring. Cooperative dynamics in a network of stochastic elements with delayed feedback. *Phys. Rev. E*, 71:036150, 2005.

[57] E. Ising. Beitrag zur Theorie des Ferromagnetismus. *Z. Phys.*, 31:253–258, 1925.

[58] E. M. Izhikevich. Neural excitability, spiking and bursting. *Int. J. Bifur. Chaos*, 10:1171, 2000.

[59] E. M. Izhikevich. Resonate-and-fire neurons. *Neural Networks*, 14:883–894, 2001.

[60] E. M. Izhikevich. *Dynamical systems in neuroscience*. The MIT Press, Cambridge, 2010.

[61] M. Januszewski and M. Kostur. Accelerating numerical solution of stochastic differential equations with CUDAstar. *Comput. Phys. Commun.*, 181:183, 2010.

[62] P. Jung, U. Behn, E. Pantazelou, and F. Moss. Collective response in globally coupled bistable systems. *Phys. Rev. A*, 46:R1709, 1992.

[63] P. Jung and G. Mayer-Kress. Spatiotemporal stochastic resonance in excitable media. *Phys. Rev. Lett.*, 74:2130–2133, 1995.

[64] H. Kager, W. J. Wadman, and G. G. Somjen. Simulated seizures and spreading depression in a neuron model incorporating interstitial space and ion concentrations. *J. Neurophysiol.*, 84(1):495–512, 2000.

[65] H. Kager, W. J. Wadman, and G. G. Somjen. The triggering of spreading depression studied with computer simulations. *J. Neurophysiol.*, 88(5):2700–2712, 2002.

[66] R. L. Kautz. Noise, chaos, and the Josephson voltage standard. *Rep. Prog. Phys.*, 59:935, 1996.

[67] J. Keener and J. Sneyd. *Mathematical Physiology*. Springer, New York, 1998.

Bibliography

[68] J. P. Keener. Propagation and its failure in coupled systems of discrete excitable cells. *SIAM J. Appl. Math.*, 47,3:556–572, 1987.

[69] V. M. Kenkre, E. W. Montroll, and M. F.Schlesinger. Generalized master equations for continuous-time random walks. *J. Stat. Phys.*, 9:45, 1973.

[70] M. Kimizuka and T. Munakata. Stochastic dynamics in systems with unidirectional delay coupling: Two-state description. *Phys. Rev. E*, 80:021139, 2009.

[71] A. Kolmogorov, I. Petrovskii, and N. Piscounov. *A study of the diusion equation with increase in the amount of substance and its application to a biological problem,.* Translated by V. M. Volosov from Bull. Moscow Univ., Math. Mech. 1, 1-25, 1937.

[72] M. Kostur, X. Sailer, and L. Schimansky-Geier. Stationary probability distributions for FitzHugh-Nagumo systems. *Fluct. Noise Lett.*, 3:L155–L166, 2003.

[73] H. A. Kramers. Brownian motion in a field of force and the diffusion model of chemical reactions. *Physica*, 7:284–304, 1940.

[74] S. Kuffler, W. Potter, and D. David. Glia in the leech central nervous system: Physiological properties and neuron-glia relationship. *J. Neurophysiol.*, 27(2):290–320, 1964.

[75] L. Kuhnert, K. I. Agladze, and V. I. Krinsky. Image processing using light-sensitive chemical waves. *Nature*, 337:244–247, 1986.

[76] Y. Kuramoto. *Lecture Notes in Phys.* Springer-Verlag, 1975.

[77] A. M. Lacasta, F. Sagués, and J. M. Sancho. Coherence and anticoherence resonance tunded by noise. *Phys. Rev. E*, 66:045105, 2002.

[78] J. S. Langer. Instabilities and pattern formation in crystal growth. *Rev. Mod. Phys.*, 52:1–28, 1980.

[79] P. Langevin. Sur la théorie du mouvement Brownien. *Compt. rend.*, 146:530, 1908.

[80] J. P. Laplante and T. Erneux. Propagation failure in arrays of coupled bistable chemical reactors. *J. Phys. Chem.*, 96, 12:4931–4934, 1992.

[81] J. Lechleiter, S. Girard, E. Peralta, and D. Clapham. Spiral calcium wave propagation and annihilation in xenopus laevis oocytes. *Science*, 252(5002):123–126, 1991.

[82] K. J. Lee and H. L. Swinney. Lamellar structures and self-replicating spots in a reaction-diffusion system. *Phys. Rev. E*, 51:1899–1915, 1995.

[83] Y. X Li and J. Rinzel. Equations for InsP3 receptor-mediated $[Ca^{2+}]_i$ oscillations derived from a detailed kinetic model: A Hodgkin-Huxley like formalism. *Journal of Theoretical Biology*, 166:461–473, 1994.

[84] K. Lindenberg, J. M. Sancho, A. M. Lacasta, and I. M. Sokolov. Dispersionless transport in a washboard potential. *Phys. Rev. Lett.*, 98:020602, 2007.

[85] B. Lindner, J. Garcia-Ojalvo, A. B. Neiman, and L. Schimansky-Geier. Effects of noise in excitable systems. *Phys. Rep.*, 392:321, 2004.

[86] B. Lindner, M. Kostur, and L. Schimansky-Geier. Optimal diffusive transport in a tilted periodic potential. *Fluct. Noise Lett.*, 1:R25, 2001.

[87] B. Lindner and L. Schimansky-Geier. Analytical approach to the stochastic FitzHugh-Nagumo system and coherence resonance. *Phys. Rev. E*, 60:7270–7276, 1999.

[88] B. Lindner and L. Schimansky-Geier. Coherence and stochastic resonance in a two-state system. *Phys. Rev. E*, 61:6103, 2000.

[89] B. Lindner and L. Schimansky-Geier. Noise-induced transport with low randomness. *Phys. Rev Lett.*, 89:230602, 2002.

[90] J. F. Lindner, B. K. Meadows, W. L. Ditto, M. E. Inchiosa, and A. R. Bulsara. Array enhanced stochastic resonance and spatiotemporal synchronization. *Phys. Rev. Lett.*, 75:3–6, 1995.

[91] M. Löcher, D. Cigna, E. R. Hunt, G. A. Johnson, F. Marchesoni, L. Gammaitoni, M. E. Inchiosa, and A. R. Bulsara. Stochastic resonance in coupled nonlinear dynamic elements. *Chaos*, 8:604, 1998.

[92] A. J. Lotka. Analytical note on certain rhythmic relations in organic systems. *Proc. Natl. Acad. Sci. U.S.A.*, 6:410, 1920.

[93] A. Malevanets and R. Kapral. Microscopic model for FitzHugh-Nagumo dynamics. *Phys. Rev. E*, 55(5):5657–5670, 1997.

[94] B. McNamara and K. Wiesenfeld. Theory of stochastic resonance. *Phys. Rev. A*, 39:4854, 1989.

[95] E. Mihaliuk, T. Sakurai, F. Chirila, and K. Showalter. Feedback stabilization of unstable propagating waves. *Phys. Rev. E*, 65:065602(R), 2002.

Bibliography

[96] C. Morris and H. Lecar. Voltage oscillations in the barnacle giant muscle fiber. *Biophysical Journal*, 35:193–213, 1981.

[97] F. Müller, L. Schimansky-Geier, V. S. Zykov, and H. Engel. Drift and diffusion in a periodic two-dimensional velocity field without attractors. *Phys. Rev. E*, 75:062101, 1994.

[98] F. Müller, L. Schimansky-Geier, V. S. Zykov, and H. Engel. Drift and diffusion in a periodic two-dimensional velocity field without attractors. *Phys. Rev. E*, 75:062101, 2007.

[99] J. D. Murray. *Mathematical biology*. Springer, Berlin, 1993.

[100] A. Neiman, L. Schimansky-Geier, A. Cornell-Bell, and F. Moss. Noise-enhanced phase synchronization in excitable media. *Phys. Rev. Lett.*, 83:4896–4899, 1999.

[101] A. B. Neiman and D. F. Russell. Models of stochastic biperiodic oscillations and extended serial correlations in electroreceptors of paddlefish. *Phys. Rev. E*, 71:061915, 2005.

[102] E. M. Nicola, L. Brusch, and M. Bär. Antispiral waves as sources in oscillatory reaction-diffusion media. *J. Phys. Chem. B*, 108 (38):14733–14740, 2004.

[103] Y. Nishiura and M. Mimura. Layer oscillations in reaction-diffusion systems. *SIAM J. Appl. Math.*, 49, 2:481–514, 1989.

[104] A. Nitzan, P. Ortoleva, J. Deutch, and John Ross. Fluctuations and transitions at chemical instabilities: The analogy to phase transitions. *J. Chem. Phys.*, 61:1056, 1974.

[105] L. L. Odette and E. A. Newman. Model of potassium dynamics in the central nervous system. *Glia*, 1(3):198–210, 1988.

[106] T. Ohira and T. Yamane. Delayed stochastic systems. *Phys. Rev. E*, 61:1247, 2000.

[107] T. Ohta, M. Mimura, and R. Kobayashi. Higher-dimensional localized patterns in excitable media. *Physica D: Nonlinear Phenomena*, 34, 1-2:115–144, 1989.

[108] E.-H. Park and D. M. Durand. Role of potassium lateral diffusion in non-synaptic epilepsy: A computational study. *J. Theor. Biol.*, 238:666–682, 2006.

[109] V. Petrov, S. K. Scott, and K. Showalter. Excitability, wave reflection, and wave splitting in a cubic autocatalysis reaction-diffusion system. *Phil. Trans. R. Soc. Lond. A*, 347:631–642, 1994.

[110] A. S. Pikovsky and J. Kurths. Coherence resonance in a noise-driven excitable systems. *Phys. Rev. Lett.*, 78:775–778, 1997.

[111] A. S. Pikovsky, M. G. Rosenblum, and J. Kurths. *Synchronization, A Universal Concept in Nonlinear Science.* Cambridge University Press, Cambridge, 2001.

[112] L. A. Pontryagin, A. Andronov, and A. Vitt. On the statistical treatment of dynamical systems. *Zh. Eksp. Teor. Fiz. (translated by J.B. Barbour and reproduced in Noise in Nonlinear Dynamics, 1989, edited by F. Moss and P.V.E. McClintock (Cambridge University Press, Cambridge)*, 3:165–180, 1933.

[113] O. V. Popovych, C. Hauptmann, and P. A. Tass. Control of neuronal synchrony by nonlinear delayed feedback. *Biol. Cybern.*, 95:69, 2006.

[114] D. E. Postnov, A. P. Nikitin, and V. S. Anishchenko. Synchronization of the mean velocity of a particle in stochastic ratchets with a running wave. *Phys. Rev. E*, 58:1662, 1998.

[115] D. E. Postnov, L. S. Ryazanova, O. S. Sosnovtseva, and E. Mosekilde. Neural synchronization via potassium signalling. *Int. J. Neural Syst.*, 16:99–109, 2006.

[116] D. E. Postnov, L. S. Ryazanova, R. A. Zhirin, E. Mosekilde, and O. S. Sosnovtseva. Noise controlled firing patterns in potassium driven neural networks,. *Int. J. Neural Syst.*, 17 (2):105–113, 2007.

[117] T. Prager. *Synchronization in periodically driven and coupled stochastic systems-A discrete state approach.* PhD thesis, Humboldt-Universität zu Berlin, 2006.

[118] T. Prager, B. Naundorf, and L. Schimansky-Geier. Coupled three-state oscillators. *Physica A*, 325:176, 2003.

[119] T. Prager and L. Schimansky-Geier. Stochastic resonance in a non-Markovian discrete state model for excitable systems. *Phys. Rev. Lett.*, 91:230601, 2003.

[120] P. Reimann. Brownian motors: noisy transport far from equilibrium. *Physics Reports*, 361:57, 2002.

[121] P. Reimann, C. V. den Broeck, H. Linke, P. Hänggi, J. M. Rubi, and A. Pérez-Madrid. Giant acceleration of free diffusion by use of tilted periodic potentials. *Phys. Rev. Lett.*, 87:010602, 2001.

[122] J. Rinzel and G. B. Ermentrout. *Methods in neuronal modeling.* The MIT Press, Cambridge, 1989.

[123] H. Risken. *The Fokker-Planck equation.* Springer Verlag, 1984.

Bibliography

[124] H. H. Rotermund, W. Engel, M. Kordesch, and G. Ertl. Imaging of spatiotemporal pattern evolution during carbon monoxide oxidation on platinum. *Nature*, 343:355–357, 1990.

[125] T. Sakurai, E. Mihaliuk, F. Chirila, and K. Showalter. Design and control of wave propagation patterns in excitable media. *Science*, 14:2009–2012, 2002.

[126] L. Schimansky-Geier, Ch. Zülicke, and E. Schöll. Domain formation due to Ostwald ripening in bistable systems far from equilibrium. *Z. Phys. B - Condensed Matter*, 84-3:433–441, 1991.

[127] F. Schlögl. Chemical reaction models for non-equilibrium phase transitions. *Z. Phys. B*, 253, 2:147–161, 1972.

[128] T. Schwalger and B. Lindner. Higher-order statistics of a bistable system driven by dichotomous colored noise. *Phys. Rev. E*, 78:210603, 2008.

[129] T. Schwalger and L. Schimansky-Geier. Interspike interval statistics of a leaky integrate-and-fire neuron driven by gaussian noise with large correlation times. *Phys. Rev. E*, 77:031914, 2008.

[130] X. Shao, Y. Wu, J. Zhang, H. Wang, and Q. Ouyang. Inward propagating chemical waves in a single-phase reaction-diffusion system. *Phys. Rev. Lett.*, 100:198304, 2008.

[131] M. B. Short, P. J. Brantingham, A. L. Bertozzi, and G. E. Tita. Dissipation and displacement of hotspots in reaction-diffusion models of crime. *PNAS*, 107-9:3961–3965, 2010.

[132] J. W. Shuai and P. Jung. Selection of intracellular calcium patterns in a model with clustered Ca^{2+} release channels. *Phys. Rev. E*, 67:031905, 2003.

[133] O. Steinbock, V. Zykov, and S. C. Müller. Control of spiral-wave dynamics in active media by periodic modulation of excitability. *Nature*, 366:322 – 324, 1993.

[134] A. Sterck, D. Koelle, and R. Kleiner. Rectification in a stochastically driven three-junction squid rocking ratchet. *Phys. Rev. Lett.*, 103:047001, 2009.

[135] J. D. Sterman. *Business Dynamics: Systems Thinking and Modeling for a Complex World*, volume 53, 4. Irwin/McGraw-Hill, 2000.

[136] R. L. Stratonovich. *Topics in the Theory of Random Noise I*. Gordon and Breach, 1962.

[137] R. L. Stratonovich. *Topics in the Theory of Random Noise II*. Cambridge University Press, Cambridge, 1967.

[138] A. B. Stundzia and C. J. Lumsden. Stochastic simulation of coupled reaction-diffusion processes. *Journal of Computational Physics*, 127:196–207, 1996.

[139] E. Syková. Extracellular K^+ accumulation in the central nervous system. *Prog. Biophys. Mol. Biol.*, 42:135, 1983.

[140] S. A. Tatarkova, W. Sibbett, and K. Dholakia. Brownian particle in an optical potential of the washboard type. *Phys. Rev. Lett.*, 91:038101, 2003.

[141] A. F. Taylor, M. R. Tinsley, F. Wang, Z. Huang, and K. Showalter. Dynamical quorum sensing and synchronization in large populations of chemical oscillators. *Science*, 323:614–617, 2009.

[142] P. Tierno, T. H. Johansen, and F. Sagués. Dynamical regimes of a paramagnetic particle circulating a magnetic bubble domain. *Phys. Rev. E*, 80:052401, 2009.

[143] Y. Timofeevaa, G. J. Lord, and S. Coombes. Dendritic cable with active spines: A modelling study in the spike-diffuse-spike framework. *Neurocomputing*, 69:1058–1061, 2006.

[144] M. R. Tinsley, A. F. Taylor, Z. Huang, and K. Showalter. Emerge of collective behavior in groups of excitable catalyst-loaded particles: Spatiotemporal dynamical quorum sensing. *Phys. Rev. Lett.*, 102:1583301, 2009.

[145] H. Treutlein and K. Schulten. Noise-induced neural impulses. *Eur. Biophys. J.*, 13:355–365, 1986.

[146] S. Trimper and K. Zabrocki. Delay-controlled reactions. *Phys. Lett. A*, 321:205, 2004.

[147] G. E. Uhlenbeck and L. S. Ornstein. On the theory of the Brownian motion. *Phys. Rev.*, 36:823–841, 1930.

[148] G. Ullah, J. R. Jr. Cressman, E. Barreto, and S. J. Schiff. The influence of sodium and potassium dynamics on excitability, seizures, and the stability of persistent states: Ii. network and glial dynamics. *J. Comput. Neurosci.*, 26:171–183, 2009.

[149] E. Ullner, A. Zaikin, J. García-Ojalvo, and J. Kurths. Noise-induced excitability in oscillatory media. *Phys. Rev. Lett.*, 91:180601, 2003.

[150] D. Sigeti und W. Horsthemke. Pseudo-regular oscillations induced by external noise. *J. Stat. Phys.*, 54, 5-6:1217–1222, 1989.

[151] V. K. Vanag and I. R. Epstein. Inwardly rotating spiral waves in a reaction-diffusion system. *Science*, 294:835–837, 2001.

Bibliography

[152] B. N. Vasiev, P. Hogeweg, and A. V. Panfilov. Simulation of dictyostelium discoideum aggregation via reaction-diffusion model. *Phys. Rev. Lett.*, 73:3173–3176, 1994.

[153] T. Verechtchaguina, L. Schimansky-Geier, and I. M. Sokolov. Spectra and waiting-time densities in firing resonant and nonresonant neurons. *Phys. Rev. E*, 70:031916, 2004.

[154] B. A. Vern, W. H. Schuette, and L. E. Thibault. $[K^+]_0$ clearance in cortex: a new analytical model. *J. Neurophysiol*, 40(5):1015–1023, 1977.

[155] V. Volterra. Variation and fluctuation of a number of individuals in animal species living together. *Translation in: R. N. Chapman: Animal Ecology*, New York: McGrew Hill:409–448, 1931.

[156] . A. White, J. T. Rubinstein, and A. R. Kay. Channel noise in neurons. *Trends Neurosci*, 23:131–137, 2000.

[157] H. R. Wilson. *Spikes, decisions, and actions: the dynamical foundations of neuroscience*. Oxford University Press, Oxford, UK, 1999.

[158] F. X. Witkowski, L. J. Leon, P. A. Penkoske, W. R. Giles, M. L. Spano, W. L. Ditto, and A. T. Winfree. Spatiotemporal evolution of ventricular fibrillation. *Nature*, 392:78–82, 1998.

[159] J. Wolf and R. Heinrich. Dynamics of two-component biochemical systems in interacting cells; synchronization and desynchronization of oscillations and multiple steady states. *BioSystems*, 43:1–24, 1997.

[160] K. Wood, C. Van den Broeck, R. Kawai, and K. Lindenberg. Critical behavior and synchronization of discrete stochastic phase-coupled oscillators. *Phys. Rev. E*, 74:031113, 2006.

[161] G. X. Yan, J. Chen, K. A. Yamada, A. G. Kleber, and P. G. Corr. Contribution of shrinkage of extracellular space to extracellular K^+ accumulation in myocardial ischemia at the rabbit. *J.Physiol.*, 490:215, 1996.

[162] S. Yanchuk and P. Perlikowksi. Delay and periodicity. *Phys. Rev. E*, 79:046221, 2009.

[163] M. K. S. Yeung and S H. Strogatz. Time delay in the Kuramoto model of coupled oscillators. *Phys. Rev. Lett.*, 82:648, 1999.

[164] C.-S. Yi, A. L. Fogelson, J. P. Keener, and C. S. Peskin. A mathematical study of volume shifts and ionic concentration changes during ischemia and hypoxia. *J. of Theor. Biol.*, 220:83, 2003.

[165] G. W. De Young and J. Keizer. A single-pool inositol 1,4,5-trisphosphate-receptor-based model for agonist-stimulated oscillations in Ca^{2+} concentration. *PNAS*, 89(9895-9899), 1992.

[166] A. M. Zhabotinsky and A. N. Zaikin. Concentration wave propagation in two–dimensional liquid–phase self–oscillating system. *Nature*, 225:535–537, 1970.

[167] V. Zykov, G. Bordyugov, H. Lentz, and H. Engel. Hysteresis phenomenon in the dynamics of spiral waves rotating around a hole. *Physica D: Nonlinear Phenomena*, 239:797–807, 2010.

i want morebooks!

Buy your books fast and straightforward online - at one of world's fastest growing online book stores! Environmentally sound due to Print-on-Demand technologies.

Buy your books online at
www.get-morebooks.com

Kaufen Sie Ihre Bücher schnell und unkompliziert online – auf einer der am schnellsten wachsenden Buchhandelsplattformen weltweit! Dank Print-On-Demand umwelt- und ressourcenschonend produziert.

Bücher schneller online kaufen
www.morebooks.de

VDM Verlagsservicegesellschaft mbH
Heinrich-Böcking-Str. 6-8　　Telefon: +49 681 3720 174　　info@vdm-vsg.de
D - 66121 Saarbrücken　　　Telefax: +49 681 3720 1749　　www.vdm-vsg.de

Printed by Books on Demand GmbH, Norderstedt / Germany